Lecture Notes in Mathematics

Edited by A. Dold and B. Eckmann

Series: Universidad Complutense de Madrid
Advisers: A. Dou and M. de Guzmán

481

Miguel de Guzmán

Differentiation of Integrals in R^n

Springer-Verlag
Berlin · Heidelberg · New York 1975

Author
Prof. Miguel de Guzmán
Facultad de Matemáticas
Universidad Complutense de Madrid
Madrid 3/Spain

Library of Congress Cataloging in Publication Data

Guzmán, Miguel de, 1936-
Differentiation of integrals in R^n

(Lecture notes in mathematics ; 481)
Bibliography: p.
Includes index.
1. Integrals, Generalized. 2. Measure theory.
I. Title. II. Series: Lecture notes in
mathematics (Berlin) ; 481.
QA3.L28 no. 481 [QA312] 510'.8s [515'.43] 75-25635

AMS Subject Classifications (1970): 26A24, 28A15

ISBN 3-540-07399-X Springer-Verlag Berlin · Heidelberg · New York
ISBN 0-387-07399-X Springer-Verlag New York · Heidelberg · Berlin

Printed in Germany
Offsetdruck: Julius Beltz, Hemsbach/Bergstr.

DEDICATED TO

MAYTE

Miguel_2 and Mayte_2

PREFACE

The work presented here deals with the local aspect of the differentiation theory of integrals. This theory takes its origin in the wellknown theorem of Lebesgue [1910]: Let f be a real function in $L^1(R^n)$. Then, for almost every $x \in R^n$ we have, for every sequence of open Euclidean balls $B(x,r_k)$ centered at x such that $r_k \to 0$,

$$\lim (1/|B(x,r_k)|) \int_{B(x,r_k)} f(y)dy = f(x) \quad \text{as} \quad k \to \infty.$$

One could think that the fact that one takes here the limit of the means over Euclidean balls instead of taking them over other type of sets contracting to the point x might well be irrelevant. It was not until about 1927 that H. Bohr exhibited an example, first published by Carathéodory [1927], showing that intervals in R^2 (i.e. rectangles with sides parallel to the axes) behave much worse than cubic intervals or circles with regard to a covering property (Vitali's lemma) that was fundamental for the result of Lebesgue. So it became a challenging problem to find out whether the replacement of Euclidean balls by intervals centered at the point x in the Lebesgue theorem would lead to a true statement or not. The first result in this direction was the so-called strong density theorem, first proved by Saks [1933], stating that if the function f is the characteristic function of a measurable set, then Ecuclidean balls can be replaced by intervals. Later on Zygmund [1934] showed that this can also be done if f is in any space $L^p(R^n)$, with $1 < p \leq \infty$, and a year later Jessen, Marcinkiewicz and Zygmund [1935] proved that the same is valid if f is in $L(1+\log^+ L)^{n-1}(R^n)$. On the other hand Saks [1934] proved that there exists a function g in $L(R^n)$ such that the Lebesgue statement is false for g if one take intervals instead of balls. The Fundamenta Mathematicae of those years, which still remains one of the main sources of information for the theory of differentiation bears testimony to the interest of many outstanding mathematicians for this subject.

One of the important products of such activity was the surprising result that, if in the Lebesgue theorem one tries to replace circles by rectangles centered at the point x then the statement is not any more true in general even if f is assumed to be the characteristic function of a measurable set. This was first observed by Zygmund as a byproduct of the construction by Nikodym [1927] of a certain paradoxical set.

Such findings prompted others to try to consider more general situations and to give some characterization of those families of sets that, like the Euclidean balls or the intervals, would permit a differentiation theorem similar to that of Lebesgue. The first attempts in this direction were the fundamental paper of Busemann and Feller [1934], giving such a characterization by means of a certain "halo" condition, and the paper by de Possel [1936], offering one in terms of a covering property.

In this way there arose the theory of differentiation, which as we shall have occasion to show, still presents many challenging open problems and has very interesting connections with other branches of analysis. In the present work I have tried to focus on some of the more fundamental aspects of the differentiation theory of integrals in R^n. In this context the theory can be presented very concretely and with a minimal amount of terminology. Many interesting open problems, whose solution will probably lead to a better undertanding of basic structures in analysis, can be stated in a way simple enough to be inmediately understood by those who just know what is a Lebesgue measurable function defined on R^2.

The differentiation theory we shall present here appears as an interaction between covering properties of families of sets in R^n, differentiation properties similar to that of the Lebesgue theorem, and estimations for an adequate extension of the wellknown maximal operator of Hardy and Littlewood. The whole book is a commentary on these three main subjects.

Chapter I is devoted to the main covering theorems that are used in the subject. Chapter II introduces the notions of a differentiation basis and of the maximal operator associated to it, and offers certain basic methods in order to obtain several useful estimations for this operator. Chapter III shows how closely related are the properties of the maximal operator and the differentiation properties of a basis. Chapters IV, V and VI explore some properties of several examples of differentiation bases, the basis of intervals, that of rectangles, and of some special sets (convex sets and unbounded star-shaped sets). Chapter VII is devoted to the possibility of obtaining covering properties starting from differentiation properties of a basis. Finally Chapter VIII contains some considerations about a particular problem in which the author has been interested.

Each chapter is divided in sections. I have tried to offer in the main body of each section just the relevant result that has been the source of inspiration for many other further developments. In the remarks at the end of each section I give information, often rather detailed, about some extensions of the theory, without trying at all to be exhaustive. In the theory we present there are still many open problems. I have stated some of them, almost always in the remarks at the end of each section. A list of them is given at the end. Some of these problems might be easy to solve, but some others seem to be rather difficult and will perhaps require fresh ideas and new techniques in our field. I hope that some of the readers will be stimulated by such problems and so the theory will be enriched with their effort. I would certainly be very grateful for any light on these problems I might receive from them. I am very happy to say that after the first version of these notes was written, in December 1.974, some of the problems proposed in them have been solved and some others have been substantially illuminated. In the appendices at the end of this work, written by A. Córdoba, R. Fefferman and R. Moriyón one can see some of the progress that has been made. I wish to thank them for having permitted me to include in these notes their results, that will be of great value for those interested in the field. Also very recently C. Hayes has solved in a very general setting the problem proposed in page 165.

I wish to thank, first of all, Prof. Antoni Zygmund for the encouragement I have received from him to write this work and for many helpful discussions on the subject. The assistance and helpful criticism of my colleagues at the University of Madrid has been invaluable. I owe particular gratitude to C. Aparicio, M.T. Carrillo, J. López, M.T. Menárguez, R. Moriyón, I. Peral, B. Rubio and M. Walias for many stimulating hours we have spent discussing the topics treated here. I also wish to thank A.M. Bruckner, C. Hayes and G. V. Welland for having read the first version of these notes and for their very helpful suggestions. Paloma Rodríguez, Isi Vázquez and Pablo Mz. Alirangües were in charge of typing and preparing these notes for publication. I thank them very much for their fine job.

Miguel de Guzmán

June 1.975

Facultad de Matemáticas

Universidad Complutense de Madrid

Madrid 3, Spain

CONTENTS

CHAPTER I

SOME COVERING THEOREMS

CHAPTER II

THE HARDY-LITTLEWOOD MAXIMAL OPERATOR

CHAPTER III

THE MAXIMAL OPERATOR AND THE DIFFERENTIATION PROPERTIES OF A BASIS

CHAPTER IV

THE INTERVAL BASIS $\mathcal{B}_2$

Page

CHAPTER V

THE BASIS OF RECTANGLES $\mathcal{B}_3$

CHAPTER VI

SOME SPECIAL DIFFERENTIATION BASES

CHAPTER VII

DIFFERENTIATION AND COVERING PROPERTIES

CHAPTER VIII

ON THE HALO PROBLEM

SOME NOTATION

For a point x in R^n, $|x|$ means the Euclidean norm of x, i.e. if x = = $(x_1, x_2, \ldots, x_n)$, then

$$|x| = (\sum_{i=1}^{n} x_i^2)^{1/2}.$$

For a set A in R^n, $|A|_e$ means the exterior Lebesgue measure of A, $\delta(A)$ the (Euclidean) diameter of A, ∂A the boundary of A. If A is measurable, $|A|$ denotes its measure, and A' denotes the complement of A.

For a sequence $\{A_k\}$ of subsets of R^n and a point $x \in R^n$, $A_k \to x$ ("A_k contracts to x") means that $x \in A_k$ for each k and $\delta(A_k) \to 0$.

For a sequence $\{r_k\}$ of real numbers and $a \in R$, $r_k \uparrow a$ ($r_k \downarrow a$) means that r_k converges increasingly (decreasingly) to a.

For a family $\mathcal{A}$ of sets of R^n, $\bigcup \{A : A \in \mathcal{A}\}$ means the set of points of R^n belonging to some set of the family $\mathcal{A}$.

CHAPTER I

SOME COVERING THEOREMS

Covering theorems of different types are extremely useful. One can think of the role the Heine-Borel theorem, the theorem of Vitali and other similar results play in Analysis.

The theory of differentation, as we shall see in this work, is a combination of covering properties and differentiation properties of certain families of sets. Among these properties, those related to covering are perhaps more basic, since in many cases only geometrical elements and not measure-theoretical ones are required in order to establish them.

In this Chapter we collect several covering results, that are useful not only in differentiation theory, but also in many other fields.

The results of the first section originate in Besicovitch [1945,1946] and A.P. Morse [1947]. They are purely geometrical and deal with certain covering properties by Euclidean balls in R^n in the case of Besicovitch and by more abstract sets in the case of Morse.

The second section presents a very useful lemma due to Whitney [1934], showing how to cover an open set in R^n by means of disjoint cubes that become smaller and smaller as one approaches the boundary.

The third section deals with the classical theorem of Vitali on the possibility of choosing from a given collection of cubes covering a set A in a certain way a disjoint sequence of such cubes that cover almost every point of A.

In all three sections we shall indicate how these theorems can be extended to be applied in many other situations which the original theorems would not cover.

1. COVERING THEOREMS OF THE BESICOVITCH TYPE.

The theorem that follows constitutes a valuable tool in the theory of differentiation and in many other fields of Analysis. Of a purely geometrical nature, it is basic in order to prove some other covering theorems, such as the one of Vitali. The original theorem of Besicovitch deals with Euclidean balls in R^n, but the following version is simpler and more elementary and one can understand through it in a clearer way the idea leading to the result.

THEOREM 1.1. Let A be a bounded set in R^n. For each $x \in A$ a closed cubic interval $Q(x)$ centered at x is given. Then one can choose, from among the given intervals $(Q(x))_{x\in A}$, a sequence $\{Q_k\}$ (possibly finite) such that:

i) The set A is covered by the sequence, i.e. $A \subset \bigcup Q_k$.

ii) No point of R^n is in more than θ_n (a number that only depends on n) cubes of the sequence $\{Q_k\}$, i.e. for every $z \in R^n$

$$\sum \chi_{Q_k}(z) \leq \theta_n.$$

iii) The sequence $\{Q_k\}$ can be distributed in ξ_n (a number that depends only on n) families of disjoint cubes.

In order to make the proof more clear we first present a more elementary version whose proof contains the main idea of the theorem.

THEOREM 1.2. Let $\{Q_k\}_{k=1}^{\infty}$ be a decreasing sequence of closed cubic intervals centered at the origin of R^n. Assume $\bigcap_{k=1}^{\infty} Q_k = \{0\}$. Let A be a bounded set of R^n. For each $x \in A$ we take a positive integer $i(x)$ and we write $Q(x) = x + Q_{i(x)}$. Then there exists a sequence $\{x_k\} \subset A$ (possibly finite) such that:

i) The set A is covered by $\{Q(x_k)\}$, i.e. $A \subset \bigcup Q(x_k)$.

ii) Each point $z \in R^n$ is in at most 2^n of the sets $Q(x_k)$, i.e. for each $z \in$

$\in R^n$, $\sum \chi_k(z) \leq 2^n$, <u>where</u> χ_k <u>is the characteristic function of</u> $Q(x_k)$.

iii) <u>The sequence</u> $\{Q(x_k)\}$ <u>can be distributed in</u> $4^n + 1$ <u>disjoint sequences</u>.

<u>Proof of Theorem</u> 1.2. We take x_1 such that $Q(x_1)$ is of maximum size. Assume that $x_1, x_2, \ldots, x_m$ have been already chosen. If

$$A - \bigcup_{k=1}^{m} Q(x_k) = \phi,$$

then the selection process is finished. Otherwise we take $x_{m+1} \in A - \bigcup_{k=1}^{m} Q(x_k)$ such that $Q(x_{m+1})$ is of maximum size. The sequence $\{Q(x_{m+1})\}$ we have thus obtained satisfies the following properties:

(1) If $i \neq j$, then $x_i \notin Q(x_j)$.

(2) The sequence $\{\delta(Q(x_k))\}$ of diameters of the sets $Q(x_k)$ is either finite or else such that $\delta(Q(x_k)) \downarrow 0$ as $k \uparrow \infty$, since the sets $x_k + \frac{1}{2} Q_{i(x_k)}$ are disjoint.

If the selection process stops, property i) is trivial. If $\{Q(x_k)\}$ is an infinite sequence and there is an $x \in A - \bigcup_{k=1}^{\infty} Q(x_k)$ then there exists j_0 such that the diameter of $Q(x)$ is bigger than that of $Q(x_{j_0})$. This would mean that x has been overlooked in our selection. Hence $A \subset \bigcup Q(x_k)$.

We now prove that each $z \in R^n$ is at most in 2^n of the sets $Q(x_k)$. To do this we draw through z n hyperplanes parallel to the n coordinate hyperplanes and consider the 2^n closed "hyperquadrants" through z determined by them. In each hyperquadrant there is at most one point x_j such that $z \in Q(x_j)$. In fact, if there were two such points x_i, x_j, the bigger of the two sets $Q(x_i)$, $Q(x_j)$ would contain the center of the smaller one and this is excluded by construction.

In order to prove iii) we fix an element $Q(x_j)$ of the sequence $\{Q(x_k)\}$. According to ii), at most 2^n members of the sequence contain a fixed vertex of $Q(x_j)$. Each cube $Q(x_k)$ with $k < j$ is of a size not smaller than that of $Q(x_j)$ and so if

$$Q(x_k) \cap Q(x_j) \neq \phi$$

with $k < j$, then $Q(x_k)$ contains at least one of the 2^n vertices of $Q(x_j)$. Hence for each $Q(x_j)$ there are at most $2^n \times 2^n$ sets of the collection

$$\{Q(x_1), Q(x_2), \ldots, Q(x_{j-1})\}$$

with non empty intersection with $Q(x_j)$. This fact allows us to distribute the sets $Q(x_k)$ in $4^n + 1$ disjoint sequences in the following way: We set $Q(x_i) \in I_i$ for $i = 1, 2, \ldots, 4^n + 1$. Since $Q(x_{4^n+2})$ is disjoint with $Q(x_{k_0})$ for some $k_0 \leq 4^n + 1$ we can set $Q(x_{4^n+2}) \in I_{k_0}$. In the same way $Q(x_{4^n+3})$ is disjoint with all sets in some I_{k^*} and so we can set $Q(x_{4^n+3}) \in I_{k^*}$, and so on. This proves iii).

<u>Proof of Theorem</u> 1.1. The natural process of selection would consist, as in the preceding proof, in choosing first the biggest possible sets and excluding the part of A already covered by them. The fact that the supremum of the diameters of the cubes may be unreachable compels us to choose the cubes <u>among</u> the biggest ones. A purely technical complication.

Let

$$a_0 = \sup\{\delta(Q(x)) : x \in A\}$$

If $a_0 = \infty$ then a single cube $Q(x)$ conveniently chosen is sufficient to cover A. If $a_0 < \infty$ we choose $Q_1 \in (Q(x))_{x\in A}$ with center $x_1 \in A$ such that $\delta(Q_1) > \frac{a_0}{2}$. Let now

$$a_1 = \sup\{\delta(Q(x)) : x \in A - Q_1\}$$

We choose Q_2 with center $x_2 \in A - Q_1$ such that $\delta(Q_2) > \frac{a_1}{2}$. And so on.

Observe that we cannot affirm that if $i \neq j$ then $x_i \notin Q(x_j)$. This certainly happens if $i > j$, but not necessarily if $i < j$. However we can always guarantee that $\frac{1}{3} Q_i \cap \frac{1}{3} Q_j = \phi$, where $\frac{1}{3} Q_k$ stands for the cube concentric with Q_k and whose size is one third of the size of Q_k. In fact, if for example $i > j$, then $x_i \notin Q_j$ and $\delta(Q_j) >$

> $\frac{1}{2}(Q_i)$, and this implies the preceding assertion. In this way we obtain i) as in the proceding proof.

The proofs of ii) and iii) are essentially the same as in the proof of Theorem 1.2. using the same remark we have used in the above paragraph in order to prove i). We omit the details and leave them as an exercise.

REMARKS.

(1) The Theorem 1.1. for unbounded A.

If A is not bounded in Theorem 1.1. but

$$\sup \{\delta(Q(x)) : x \in A\} = M < \infty$$

the theorem 1.1. is still valid changing conveniently the constants θ_n and ξ_n. To show this, it is sufficient to partition R^n in disjoint cubic intervals I_i of side length M and to apply theorem 1.1. to the intersection of A with each one of these cubic intervals I_i. The sequences $\{Q_k^i\}_{k\geq 1}$ corresponding to the different sets $A \cap I_i$ cannot overlap too much. We leave the details as an exercise.

(2) In Theorem 1.1. the intervals Q(x) can be assumed not closed.

Observe that the fact that Q(x) is closed is rather irrelevant for the proof of the theorem. Assuming that Q(x) is a cubic interval centered at x with part of the boundary changes only the size of the constants θ_n, ξ_n.

(3) The original theorem of Besicovitch [1945].

The theorem can be stated in the following way.

Let A be a bounded set of R^n. For each $x \in A$ a closed Euclidean ball $\overline{B}(x,r(x))$ with center x and radius r(x) is given. Then one can select from among $(\overline{B}(x,r(x)))_{x\in A}$ a sequence of balls $\{B_k\}$ satisfying properties i), ii), iii) of Theorem 1.1. (The constants Θ_n, ξ_n, of course, are not the same as in that theorem).

The proof follows the same pattern of that of Theorem 1.1. and is left as an exercise. For ii), instead of dealing with "hyperquadrants" in order to exclude that $z \in R^n$ is in many cubes, one proceeds in the following way. If the balls selected according the scheme of the proof of Theorem 1.1. are $\{B_k\}$, one proves that the balls $\{\frac{1}{16} B_k\}$ are disjoint, and so z is at most in one such set $\frac{1}{16} B_k$. Now if $z \in B_j - \frac{1}{16} B_j$ and j is the smallest index with this property, one draws the solid cone V with vertex z projecting the ball $\frac{1}{100} B_j$ and one proves that inside this solid cone there cannot be many centers of balls B_k such that $z \in B_k$. The whole space R^n is filled with not too many solid cones of the type of V and this concludes the proof of ii). Property iii) is obtained from ii) following the same process as in the theorem with the changes indicated here for the proof of ii).

(4) A general theorem of A.P. Morse [1947].

When one performs the proof of the theorem of Besicovitch according the lines indicated in the preceding remark, one observes that the same proof is valid in the following more general situation .

Let A be a bounded set of R^n. For each $x \in A$ a set $H(x)$ is given satisfying the two following properties: (a) there exists a fixed number $M > 0$, independent of x, and two closed Euclidean balls centered at x, $\overline{B}(x,r(x))$ and $\overline{B}(x,Mr(x))$, such that $\overline{B}(x,r(x)) \subset H(x) \subset \overline{B}(x,Mr(x))$; (b) for each $z \in H(x)$, the set $H(x)$ contains the convex hull of the set

$$\{z\} \cup \overline{B}(x,r(x)).$$

Then one can select from among $(H(x))_{x\in A}$ a sequence H_k satisfying i), ii), iii) of Theorem 1.1. (The constants Θ, ξ, depend now on n and M).

This theorem comprises the theorem of Besicovitch and Theorem 1.1. and is due to A.P. Morse [1947] independently of Besicovitch.

It would be interesting to have an estimate of the constants $\Theta(n,M)$, $\xi(n,M)$ of the theorem of Morse. For n = 1 one obtains easily $\Theta(1,M) \leq cM \log M$ for large M. One can see Guzmán [1970].

(5) In Theorem 1.1. the cubic intervals $Q(x)$ can be substituted by comparable intervals.

It is not difficult to observe that the proof of Theorem 1.1. is also valid without any essential change in the following more general situation.

Let A be a bounded set of R^n. For each $x \in A$ a closed interval $R(x)$ centered at x is given. Assume that for every two points x_1, x_2 of A the intervals $R(x_1)$, $R(x_2)$ are comparable, i.e. when translated to be concentric one is contained in the other. Then from among $(R(x))_{x\in A}$ one can select a sequence $\{R_k\}$ for which i), ii), iii) of Theorem 1.1. hold (The constants, Θ_n, ξ_n, depend only on n).

This theorem is not included in that of Morse. It was obtained for the first time as a particular case of a theorem of the Besicovitch type for a product of metric spaces. It can be seen in Guzmán [1970]*.

(6) The cubes in Theorem 1.1. have to be more or less centered at the corresponding points.

According to the remark (4) it is not necessary to have the cubes exactly centered at the corresponding points. However Theorem 1.1. is not valid anymore if x can be in the boundary of $Q(x)$ or arbitrarily close to it. To see this it suffices to take $A = (0,1]$ in R^1 and for each $x \in A$ the interval $Q(x) = [x,x+1]$. Clearly one cannot have at the same time i) and iii).

(7) An application. A generalization of Sard's theorem.

In the sections to follow we shall see several applications of these covering theorems of Besicovitch type. Certain useful generalizations of the classical theorems of Whitney and Vitali we shall present in the following sections are easy consequences of the theorems of this one. Here we shall present the following generalization of the well-known theorem of Sard on the critical points of a function. This extension is due to Guzmán [1972]*.

Let G be an open set of R^n. Let X be any set where there is an exterior measure ν defined on the subsets of X. Let $f : G \to X$ be an arbitrary function. A point x of G will be called critical point of f if there is sequence $\{Q_k(x)\}$ of open cubic intervals centered at x contracting to x, i.e. with $\delta(Q_k(x)) \to 0$, such that

$$\frac{\nu(f(Q_k(x))}{|Q_k(x)|} \to 0$$

as $k \to \infty$.

Let C be the set of critical points of f. Then $\nu(f(C)) = 0$.

The proof is easy. Let S any bounded subset of C. We shall prove that $\nu(f(S)) = 0$. Since $C = \bigcup S_k$, with S_k bounded, we get

$$\nu(f(C)) = \nu(f(\bigcup S_k)) = \nu(\bigcup f(S_k)) \leq \sum \nu(f(S_k)) = 0$$

In order to prove that $\nu(f(S)) = 0$ we take $\varepsilon > 0$ and 0 open and bounded so that $S \subset 0$.

If $x \in S$ we can choose an open cubic interval $Q(x)$ centered at x so that $Q(x) \subset 0$ and

$$\nu(f(Q(x))) < \varepsilon\ |Q(x)|$$

Applying Theorem 1.1. we choose a sequence $\{Q_k\}$ of such cubes so that $S \subset \subset \bigcup Q_k \subset 0$, and

$$\sum \chi_{Q_k} \leq \theta_n,\ \nu(f(Q_k)) \leq \varepsilon |Q_k|$$

Hence

$$\nu(f(S)) \leq \nu(f(\bigcup_k Q_k)) = \nu(\bigcup_k f(Q_k)) \leq \sum_k \nu(f(Q_k)) \leq$$

$$\leq \varepsilon \sum_k |Q_k| = \varepsilon \sum \int \chi_{Q_k}(y)dy = \varepsilon \int_{\bigcup Q_k} (\sum_k \chi_{Q_k}(y))dy \leq$$

$$\leq \varepsilon \theta_n |\bigcup_k Q_k| \leq \varepsilon \theta_n |0|$$

Since ε is arbitrarily small, $\nu(f(S)) = 0$ and the theorem is proved.

It is not difficult to see that if $X = R^n$, ν is the exterior Lebesgue measure in R^n and $f \in C^1(R^n)$, then $\left[\det \frac{\partial f}{\partial x}\right]_{x=x_0} = 0$ implies that x_0 is a critical point in the sense of the theorem. So the theorem of Sard results as an easy consequence of this theorem.

2. COVERING THEOREMS OF THE WHITNEY TYPE.

Whitney [1934] introduced a type of covering theorem which has proved later extremely useful in Analysis and Geometry. We shall study here in the first place the original version of Whitney. The theorem refers to the posibility of partitioning any open set G of R^n with non empty boundary, $\partial G \neq \phi$, into a disjoint sequence $\{Q_j\}$ of half-open cubic intervals so that the diameter of the sets Q_j is comparable with the distance of Q_j to the boundary of G. A simple geometric consideration shows that such partition is impossible if we try to use, for example, Euclidean balls instead of cubic intervals. It is also clear that Whitney's theorem and its proof are extremely geometric, so that when one tries to obtain a more abstract version of it, for instance in a general metric space, one cannot guess what should replace the cubic intervals in the abstract formulation.

The second theorem presented here, due to Coifman and Guzmán [1970] shows that Whitney's theorem is also valid for Euclidean balls in R^n if one gives up the requirement that the balls be disjoint and replaces it for a condition of uniformly bounded overlap. In the proof of this theorem one can easily realize that the space

R^n can be replaced by any metric space whose balls satisfy a certain "homogeneity" condition. In this way one obtains Theorem 2.3.

THEOREM 2.1. *Let G be any open set of R^n with non empty boundary ∂G. Then one can write $G = \bigcup_{j=1}^{\infty} Q_j$, where the sets Q_j are half-open cubic intervals (i.e. sets of the type*

$$Q(x,a) = \{z \in R^n : x_i - a \leq z_i < x_i + a,\ i=1,2,\ldots,n\})$$

disjoint and such that for each j one has

$$1 \leq \frac{d(Q_j,\partial G)}{\delta(Q_j)} < 3$$

Proof. In the proof of the theorem we shall make use of the *dyadic cubes* of R^n, that we now introduce. We first consider in R^n the family D_1 of all half-open cubic intervals (open to the right and closed to the left) of side-length equal to 1 having vertices al all points of R^n with integral coordinates. We subject now D_1 to a homothecy with center 0 and ratio 2^k for $k \in Z$ and so obtain D_k. Each cube of D_j is the union of 2^n disjoint cubes of D_{j-1}. The cubes of D_j have side-length 2^j. It is clear that if $Q_1 \in D_j$ and $Q_2 \in D_k$ with $j \leq k$, then either $Q_1 \cap Q_2 = \phi$ or else $Q_1 \subset Q_2$. Each D_j contains a countable number of cubes and the family $D = \bigcup_{k \in Z} D_k$ of all dyadic intervals of R^n is also denumerable.

For the proof of the theorem we can assume, performing a translation, if necessary, that $0 \in \partial G$. For each $x \in G$ there is at least a dyadic cube $Q(x)$ such that $x \in Q(x)$ and $0 \in \partial Q(x)$. Hence, if we consider the sequence of all dyadic intervals containing x and contracting to x, we can select a tail of this sequence $\{T_k\}_{k=1}^{\infty}$ such that

$$d(T_1,\partial G) = 0,\ T_2 \subset G,\ d(T_2,\partial G) > 0$$

and the cubes T_k belong to sucessive dyadic families D_{j_k} (i.e. if $T_1 \in D_{j_1}$, then $T_2 \in D_{j_1-1}$, $T_3 \in D_{j_1-2}$, ...).

Since $d(x,\partial G) > 0$ and $\{T_k\}$ contracts to x, we have

$$\frac{d(T_k,\partial G)}{\delta(T_k)} \to \infty \quad \text{as} \quad k \to \infty$$

We take the first T_k of the sequence $\{T_k\}$ such that

$$\frac{d(T_k,\partial G)}{\delta(T_k)} \geq 1$$

Let this cube be T_h. Observe that $h > 1$ since $d(T_1,\partial G) = 0$.

We have

$$\frac{d(T_{h-1},\partial G)}{\delta(T_{h-1})} < 1$$

and so

$$1 \leq \frac{d(T_h,\partial G)}{\delta(T_h)} \leq \frac{d(T_{h-1},\partial G) + \delta(T_h)}{\delta(T_h)} = 2\,\frac{d(T_{h-1},\partial G)}{\delta(T_{h-1})} + 1 < 3$$

Therefore, for each $x \in G$ we have a dyadic cube $Q^*(x)$ such that $x \in Q^*(x) \subset G$ and

$$1 \leq \frac{d(Q^*(x),\ G)}{(Q^*(x))} \leq 3$$

We take all cubes $Q^*(x)$, $x \in G$. They satisfy all conditions of the theorem except that of being disjoint. We now try to select from $A = (Q^*(x))_{x\in G}$ a sequence of disjoint cubes that still covers G. Let

$$\dots \subset Q^*(x_{-2}) \subset Q^*(x_{-1}) \subset Q^*(x_0) \subset Q^*(x_1) \subset \dots$$

a chain of cubes of A such that all its members are different. We have, for each $j > 0$

$$1 \leqslant \frac{d(Q^*(x_j), \partial G)}{\delta(Q^*(x_j))} \leqslant \frac{d(x_0, \partial G)}{\delta(Q^*(x_j))}$$

If the chain has an infinite number of members to the right of $Q^*(x_0)$ then, since all of them are different, $\delta(Q^*(x_j)) \to \infty$ for $j \to \infty$ and this contradicts

$$1 \leqslant \frac{d(x_0, \partial G)}{\delta(Q^*(x_j))} \quad \text{for } j > 0.$$

Hence the chain has a maximal element. We take the maximal element of each one of the possible chains of A. This family $\{Q_k\}$ of dyadic intervals is denumerable, since D is so, is disjoint, because they are dyadic and maximal in the sense explained, and cover G, since they cover each $Q^*(x)$ of A. So this family satisfies all requirements of the theorem.

If we content ourselves with representing the open set G, with $\partial G \neq \phi$, by means of a sequence of open cubes $\{Q_k\}$, this time not disjoint, but with a uniformly bounded overlap, and such that $d(Q_k, \partial G) = 3\delta(Q_k)$, we can do it easily by means of the theorems of Section 1 in the way indicated in the theorem that follows. This form of the Whitney theorem is already sufficient for a good part of the applications of Whitney's theorem and lends itself easily to generalizations.

THEOREM 2.2. *Let G be any open set of R^n with empty boundary, $\partial G \neq \phi$. Then one can set $G = \bigcup_{k=1}^{\infty} Q_k$, where the sets Q_k are open cubic intervals cuch that*

$$d(Q_k, \partial G) = 3\delta(Q_k)$$

and for each $z \in R^n$

$$\sum_{k=1}^{\infty} \chi_{Q_k}(z) \leqslant \alpha_n$$

α_n being a constant that depends only on n.

Proof. Assume first that G is bounded. For each $x \in G$ we take an open cubic interval Q(x) centered at x such that

$$d(Q(x), \partial G) = 3\delta(Q(x)).$$

A simple application of Theorem 1.1. gives us the theorem with $\alpha_n = \Theta_n$.

If G is not bounded then some merely technical complications arise. Fix k e e Z and let H_k be the set of all points x of G such that

$$2^k < d(x, \partial G) \leq 2^{k+1}$$

For each x e H_k we take an open cubic interval centered at x such that

$$d(Q(x), \partial G) = 3\delta(Q(x)).$$

So we have

$$\delta(Q(x)) \leq \frac{2^{k+1}}{3}$$

and we can apply to H_k Theorem 1.1. (cf. Remark (1) of Section 1), obtaining $\{Q_j^k\}_{j\geq 1}$ such that

$$H_k \subset \bigcup_{j\geq 1} Q_j^k , \quad \sum_{j\geq 1} \chi_{Q_j^k} \leq \Theta$$

Where Θ depends only on n.

Let us now observe that the sets of the sequence $\{Q_j^k\}_{j\geq 1}$ covering H_k and those of $\{Q_j^{k+4}\}_{j\geq 1}$ covering H_{k+4} are disjoint. In fact, assume that z e $Q_j^k \cap Q_s^{k+4}$, a is the center of Q_j^k and b is that of Q_s^{k+4}. Since z e Q_j^k we can write

$$d(z,\partial G) \leq d(z,a) + d(a,\partial G) \leq \frac{2^{k+1}}{3} + 2^{k+1}$$

Since $z \in Q_s^{k+4}$, we have

$$d(z,\partial G) \leq d(b,\partial G) - d(b,z) > 2^{k+4} - \frac{2^{k+4+1}}{3}$$

This implies $2^3 < 4$. Hence $Q_j^k \cap Q_s^{k+4} = \phi$ for each s, j.

Therefore if we take the cubes $\{Q_j^k\}_{k\in Z,\ j=1,2,\dots}$ we have a family that satisfies all requirements of the theorem.

Observe that the preceding argument does not depend on the particular geometric structure of the cubic intervals, but rather on the possibility of applying a property of the Besicovitch type to the sets one considers. One can therefore substitute in Theorem 2.2. cubic intervals by Euclidean balls or by other kind of sets such as the ones considered in the remarks of Section 1.

The following theorem is based on ideas similar to those exploited in Theorem 2.2. and its proof can be performed along the same line as there. We leave it as an interesting exercise.

THEOREM 2.3. *Let (X,ρ) be a metric space with the following "homogeneity" condition: For each $x \in X$ and for each $r > 0$ there are at most N (an absolute constant independent of x and r) points $\{x_i\}$ in*

$$B(x,r) = \{z \in X : \rho(z,x) < r\}$$

such that

$$\rho(x_i,x_j) \geq \frac{r}{2} \quad \text{for } i \neq j$$

Then every open set G of X, with $\partial G \neq \phi$, can be represented in the form

$$G = \bigcup_{k=1}^{\infty} B(x_k, r_k)$$

<u>so that each</u> $z \in G$ <u>is in at most</u> $9N$ <u>of the sets</u> $B(x_k,r_k)$ <u>and for each</u> k <u>we have</u>

$$r_k \leq d(B(x_k,r_k),\partial G) \leq 4\, r_k.$$

<u>REMARKS.</u>

(1) <u>An application. A lemma of Calderón and Zygmund.</u>

There are many applications of the theorems of this section beginning with those found by Whitney [1934] himself. In Chapter II we shall find another one related to a converse of the maximal inequality of Hardy and Littlewood. Here we shall present a proof of a wellknown lemma of Calderón and Zygmund [1952] that follows the same pattern as that of Theorem 1.1.

<u>Let</u> f <u>be a function in</u> $L^1(R^n)$, $f \geq 0$ <u>and</u> λ <u>a positive real number. Then there existe a countable disjoint family (possibly empty) of half-open cubic intervals</u> $\{Q_k\}_{k\geq 1}$ <u>such that for each</u> k

$$\lambda < \frac{1}{|Q_k|}\int_{Q_k} f \leq 2^n \lambda$$

<u>and</u> $f(x) \leq \lambda$ <u>at almost every</u> $x \notin \bigcup Q_k$.

For the proof we shall assume here the following property of the dyadic intervals of R^n, that will be proved in the next section: At almost every $x \in R^n$,

$$\lim_{k\to\infty} \frac{1}{|Q_k(x)|}\int_{Q_k(x)} f(y)\, dy = f(x)$$

where $\{Q_k(x)\}_{k=1}^{\infty}$ is any sequence of dyadic cubic intervals containing x and contracting; to x.

From the above property it is clear that if

$$\Lambda = \{x \in R^n : f(x) > \lambda\}$$

then at almost every $x \in A$ we have, for any sequence of dyadic cubic intervals $\{Q_k(x)\}_{k=1}^{\infty}$ containing x and contracting to x

$$\frac{1}{|Q_k(x)|}\int_{Q_k(x)} f > \lambda$$

if k is sufficiently big. Let us call A^* the subset of A where this happens. Since

$$\frac{1}{|Q_k(x)|}\int_{Q_k(x)} f \leq \frac{||f||_1}{|Q_k(x)|}$$

is smaller than λ for $|Q_k(x)|$ sufficiently big, it is clear that considering all dyadic intervals containing x, there is one $Q^*(x)$ such that

$$\frac{1}{|Q^*(x)|}\int_{Q^*(x)} f \leq \lambda$$

while for the dyadic interval $Q(x)$ containing x and half the size of $Q^*(x)$ we have

$$\frac{1}{|Q(x)|}\int_{Q(x)} f > \lambda$$

To each $x \in A^*$ we assign $Q(x)$. Observe that $|Q(x)| < \frac{1}{\lambda}\,||f||_1$ for each $x \in A^*$. From the collection $(Q(x))_{x\in A^*}$ we choose first one of the biggest possible sets and take away the ones which are contained in it. We proceed in the same way with the remaining ones and so we obtain a sequence $\{Q_k\}$ of dyadic cubic intervals such that if Q_k^* is the set $Q^*(x)$ corresponding to Q_k we have

$$\lambda < \frac{1}{|Q_k|}\int_{Q_k} f = \frac{2^n}{|Q_k^*|}\int_{Q_k} f \leq \frac{2^n}{|Q_k^*|}\int_{Q_k^*} f \leq 2^n\,\lambda$$

It is quite clear that $A^* \subset \bigcup Q_k$ since the sets $Q(x)$ cover A^*, and so, for almost every $x \notin \bigcup Q_k$ we have $f(x) \leq \lambda$.

This proof of the Calderón-Zygmund lemma is also valid in many other situations where the same kind of argument can be applied such as those considered in

Theorems 2.2. and 2.3. Of course, disjointness has to be substituted by uniformly bounded overlap as we have done there. This generalization has been introduced by Coifman and Guzmán [1970] in order to treat singular integral operator in more abstract contexts. This theorem has been further exploited by Coifman and Weiss [1971] in situations similar to the ones of the next remark.

(2) A "homogeneous" metric space.

Let (X,ρ) be a metric space and assume that there is in X a nontrivial Borel measure with the following property: For each $x \in X$ and $R > 0$ one has

$$\mu(B(x,R)) \leq c\mu(B(x,\tfrac{R}{2}))$$

$B(x,R)$ being the open ball centered at x and of radius R, and c being an absolute constant independent of x and R.

Then the space X satisfies the homogeneity property of Theorem 2.3.

In fact, assume that $B(y,r)$ contains $\{y_1, y_2, \ldots, y_m\}$ points such that $\rho(y_i,y_j) > \frac{r}{2}$ for $i \neq j$. Then $B(y_i, \frac{r}{2}) \subset B(y, 2r)$ for each $i = 1, 2, \ldots, m$ and

$$B(y_i,\tfrac{r}{4}) \cap B(y_j,\tfrac{r}{4}) = \phi$$

for $i \neq j$. So we have

$$\mu(B(y,2r)) \geq \sum_{i=1}^{m} \mu(B(y_i,\tfrac{r}{4}))$$

On the other hand $B(y_i,2r) \supset B(y,r)$ and so by the above condition

$$\mu(B(y,r)) \leq \mu(B(y_i,2r)) \leq c^3\mu(B(y_i,\tfrac{r}{4}))$$

Hence

$$c\mu(B(y,r)) \geq \mu(B(y,2r)) \geq \sum_{i=1}^{m} \mu(B(y_i,\tfrac{r}{4})) \geq \frac{m}{c^3}\,\mu(B(y,r))$$

and so, if $\mu(B(y,r)) \neq 0$, $m \leq c^4$.

But the condition $\mu(B(y,r)) > 0$ is a non triviality condition for μ, since if there is a single ball $B(z,\lambda)$, $\lambda > 0$, such that $\mu(B(z,\lambda)) = 0$, then in virtue of the homogeneity condition, we easily gen that $\mu(B(u,R)) = 0$ for any other ball.

Other examples and applications can be seen in Coifman and Weiss [1971].

(3) A covering problem of Fefferman.

Perhaps this is the best place to present an interesting problem proposed by C. Fefferman (oral communication) with applications in Harmonic Analysis. It consists in establishing whether the following property is true or not: For every open bounded set G of R^2 there is a sequence $\{R_k\}$ of open intervals such that:

i) $G \subset \bigcup R_k$

ii) $\sum |R_k| \leq c|G|$, c being an absolute constant independent of G.

iii) For each interval $R \subset G$ there is an interval R_h in the sequence $\{R_k\}$ such that $R \subset R_h$.

If one changes "interval" everywhere into "cubic interval" in the above problem the resulting statement is true. Its proof can be obtained in the following way. Consider the collection $(Q_\alpha)_{\alpha \in A}$ of all open cubic intervals contained in G. Let

$$a_0 = \sup \{|Q_\alpha| : \alpha \in A\}$$

and choose from $(Q_\alpha)_{\alpha \in A}$ a cube Q_1^* such that $|Q_1^*| > \frac{a_0}{4}$. Let Q_1 be the cube concentric with Q_1^* and such that $|Q_1| = 64\,|Q_1^*|$. From $(Q_\alpha)_{\alpha \in A}$ exclude those cubes having non empty intersection with Q_1^*. Let $(Q_\alpha)_{\alpha \in A_1}$ the remaining collection of cubes. We proceed with $(Q_\alpha)_{\alpha \in A_1}$ in the same way obtaining Q_2, and so on. It is not difficult to show that $\{Q_k\}$ satisfies the requirements i), ii), iii).

Carleson has recently proved that the above property is not true with the generality it has been formulated. However it still seems of interest to try to characterize in some way classes of open sets admitting a good covering by intervals such as the one required by Fefferman.

3. COVERING THEOREMS OF THE VITALI TYPE.

The most classical covering theorem in differentiation theory is that of Vitali [1908], which has traditionally been the tool to obtain the Lebesgue differentiation theorem in R^n.

In its original form the theorem of Vitali refers to closed cubic intervals and the Lebesgue measure. Later on Lebesgue [1910] and others gave it a less rigid geometric form replacing cubes by other sets "regular" with respect to cubes, keeping always the restriction to the Lebesgue measure. This restriction originates in the type of proof of the theorem, essentially that given by Banach [1924], which requires that homothetic sets have comparable measures.

Also Caratheodory's proof [1927] is based on this property although it is a little different.

Besicovitch [1945,1946] and A.P. Morse [1947] were the first in obtaining similar covering lemmas for more general measures in order to prove differentiability properties analogous to that of the Lebesgue theorem. Their method consists in using the geometric considerations leading to the results of Section 1.

In this section we first present the classical version of Vitali's theorem. Since Banach's proof is the one appearing in most textbooks in measure theory we offer here an easier proof which is more easily open to the generalizations we give later.

Based on the theorems of Section 1 we obtain then a more powerful theorem of the Vitali type, valid for a general measure.

In the remarks at the end of this section we indicate how the above theorems and their proofs keep their validity in more general situations.

THEOREM 3.1. Let A be an arbitrary set of R^n. For each $x \in A$ a sequence $\{Q_k(x)\}_{k=1}^{\infty}$ is given of closed cubic intervals centered at x and contracting to x. Then it is possible to choose from $T = (Q_k(x))_{x \in A,\ k=1,2,\dots}$ a disjoint sequence $\{S_k\}$ such that

$$|A - \bigcup S_k| = 0$$

Proof. Assume first that A is in the interior of the unit cubic interval Q with a vertex at the origin and another at the point (1, 1, ..., 1). We can assume, without loss of generality that all cubes of T are contained in the interior of Q.

We choose first $S_1 \in T$ such that

$$|S_1| \geq \frac{1}{2} \sup \{|I| : I \in T\}.$$

Then we choose $S_2 \in T$ such that $S_2 \cap S_1 = \phi$ and

$$|S_2| \geq \frac{1}{2} \sup \{|I| : I \in T,\ I \cap S_1 = \phi\}$$

Then we select $S_3 \in T$ such that $S_3 \cap (\bigcup_{i=1}^{2} S_i) = \phi$ and

$$|S_3| \geq \frac{1}{2} \sup \{|I| : I \in T,\ I \cap (\bigcup_{i=1}^{2} S_i) = \phi\}$$

and so on. This selection process is finite or infinite. If finite, then clearly $A \subset \bigcup_i S_i$. Assume that it is infinite, giving us the sequence $\{S_k\}_{k=1}^{\infty}$.

Let us prove first that for each $I \in T$ we necessarily have

$$I \cap (\bigcup_{k=1}^{\infty} S_k) \neq \phi.$$

In fact, if not, there exists S e T such that

$$S \cap (\bigcup_{k=1}^{\infty} S_k) = \phi$$

and

$$|S| > \frac{1}{2} \sup \{|I| : I \in T,\ I \cap (\bigcup_{k=1}^{\infty} S_k) = \phi\}$$

Since the sets S_k are disjoint and contained in Q, we have $|S_k| \to 0$ as $k \uparrow \infty$ and so there is a j_0 such that $|S_{j_0}| < \frac{1}{2}|S|$ Taking account the form we have chosen S_{j_0}, we obtain

$$\frac{1}{2}|S| > |S_{j_0}| \geq \frac{1}{2} \sup \{|I| : I \in T,\ I \cap (\bigcup_{k=1}^{j_0-1} S_k) = \phi\} \geq \frac{1}{2}|S|$$

This contradiction proves that for each I e T one has

$$I \cap (\bigcup_{k=1}^{\infty} S_k) \neq \phi.$$

In order to prove now that

$$|A - \bigcup_{k=1}^{\infty} S_k| = 0$$

it suffices to show that for every given $\varepsilon > 0$ there exists h such that

$$|A - \bigcup_{k=1}^{h} S_k|_e \leq \varepsilon.$$

Given any $\eta > 0$ we choose h such that

$$\sum_{h+1}^{\infty} |S_k| \leq \eta.$$

Now we can write

$$A - \bigcup_{k=1}^{h} S_k \subset \bigcup\{S : S \in T,\ S \cap (\bigcup_{k=1}^{h} S_k) = \phi\} =$$

$$= \bigcup\{S : S \in T,\ S \cap (\bigcup_{k=1}^{h} S_k) = \phi,\ S \cap (\bigcup_{k=h+1}^{\infty} S_k) \neq \phi\} =$$

$$= \bigcup_{j=h}^{\infty} (\bigcup\{S : S \in T,\ S \cap (\bigcup_{k=1}^{j} S_k) = \phi,\ S \cap S_{j+1} \neq \phi\}) \subset$$

$$\subset \bigcup_{j=h}^{\infty} (\bigcup\{S : S \in T,\ |S| \leq 2|S_{j+1}|,\ S \cap S_{j+1} \neq \phi\})$$

The first inclusion is clear, because, since $\bigcup_{1}^{h} S_k$ is compact, for each $x \in A - \bigcup_{1}^{h} S_k$ there is a $Q_j(x) \in T$ such that

$$Q_j(x) \cap (\bigcup_{1}^{h} S_k) = \phi.$$

The equality coming next is also clear by virtue of the property we have proved that for every $S \in T$ we have

$$S \cap (\bigcup_{1}^{\infty} S_k) \neq \phi.$$

The next equality is trivial and the final inclusion holds because of the selection process for the sets S_k.

So we have

$$|A - \bigcup_{1}^{h} S_k|_e \leq \sum_{j=h+1}^{\infty} 9^n |S_j| \leq 9^n \eta$$

Making $9^n \eta \leq \varepsilon$ we obtain the result.

If A is not inside Q we apply the result we have obtained to the intersection of A with the interior of each one of the cubes of R^n with side-length 1 and vertices at the points with integral coordinates.

The proof we have presented lends itself to interesting generalizations that we shall present later on in the final remarks of this section. The theorem itself, as we shall see now, is valid for a general measure. To this theorem we arrive by means of the result 1.1.

THEOREM 3.2. Let μ be any nonnegative measure in R^n that is defined on the Lebesgue measurable sets of R^n. Let μ^* be the exterior measure associated to μ, i.e. for any $P \subset R^n$,

$$\mu^*(P) = \inf \{\mu(H) : H \supset P, \ H \ \mu\text{-measurable}\}.$$

For each $x \in A$ a sequence $\{Q_k(x)\}_{k=1}^{\infty}$ is given of closed cubic intervals centered at x and contracting to x. Then one can choose from among $T=(Q_k(x))_{x\in A, k=1,2,...}$ a sequence $\{S_k\}$ such that

$$\mu^*(A - \bigcup_k S_k) = 0.$$

Proof. For each $x \in A$ we choose from $\{Q_k(x)\}_{k=1}^{\infty}$ an interval Q(x) with diameter less than or equal to 1. We apply Theorem 1.1. (cf. Remark (1) of Section 1) obtaining a sequence $\{Q_k\} \subset T$ such that

i) $A \subset \bigcup Q_k$.

ii) $\sum \chi_{Q_k}(y) \leq \theta$ for each $y \in R^n$.

iii) $\{Q_k\}$ can be distributed into ξ disjoint sequences $\{Q_k^1\}, \{Q_k^2\}, \ldots, \{Q_k^{\xi}\}$.

Because of iii), for at least one of these sequences, for example $\{Q_k^1\}$, holds

$$\mu^*(A \cap (\bigcup_k Q_k^1)) > \frac{1}{\xi} \mu^*(A).$$

In fact, otherwise we sould have

$$\mu^*(A) \leq \sum_{j=1}^{\xi} \mu^*(A \cap (\bigcup_k Q_k^j)) < \mu^*(A)$$

and this is contradictory.

From $\{Q_k^1\}$ we take a finite sequence that we call $\{S_k\}_{k=1}^{h_1}$ such that

$$\mu^*(A \cap (\bigcup_1^{h_1} S_k)) > \frac{1}{2\xi}\, \mu^*(A)$$

We now take a μ-measurable set M such that $M \supset A$ and $\mu^*(A) = \mu(M)$. We then have

$$\mu(M \cap (\bigcup_1^{h_1} S_k)) \geq \mu^*(A \cap (\bigcup_1^{h_1} S_k)) > \frac{1}{2\xi}\, \mu(M)$$

Hence

$$\mu^*(A - \bigcup_1^{h_1} S_k) \leq \mu(M - \bigcup_1^{h_1} S_k) < (1 - \frac{1}{2\xi})\, \mu^*(A) = \alpha\mu^*(A)$$

For each $x \in A - \bigcup_1^{h_1} S_k = A_1$ we take now an interval $Q(x) \in T$ with diameter less than or equal to 1 such that

$$Q(x) \cap (\bigcup_1^{h_1} S_k) = \phi$$

and we proceed with A_1 as we have done with A, obtaining now $\{S_k\}_{k=h_1+1}^{h_2}$ such that

$$\mu^*(A - \bigcup_1^{h_2} S_k) = \mu^*(A_1 - \bigcup_{h_1+1}^{h_2} S_k) < \alpha\mu^*(A_1) < \alpha^2\mu^*(A)$$

In this way we obtain $\{S_k\}$ satisfying all requirements of the theorem.

REMARKS.

(1) The theorem of Vitali with respect to the Lebesgue measure and comparable convex sets.

The proof of Theorem 3.1. we have given possesses a rather flexible structure. The reader is invited to prove, following the line of that proof, the following interesting result.

Let A be a set in R^n. For each $x \in A$ a sequence $\{K_k(x)\}$ is given of compact convex sets containing x and such that $\delta(K_k(x)) \to 0$ as $k \to \infty$. Assume that the collection given $V = (K_k(x))_{x\in A,\ k=1,2,\dots}$ is such that if $T_1, T_2 \in V$ then there is a translation of one of them that puts it inside the other. Then it is possible to select from V a disjoint sequence $\{S_k\}$ such that $|A - \bigcup S_k| = 0$.

The only significant change that is needed is the following easy geometrical observation. If $T_1, T_2 \in V$ and $|T_1| > \frac{|T_2|}{2}$ then, if one takes the union H of all the sets obtained by translating T_2 that have nonempty intersection with T_1, one has

$$|H| \leq 9^n \, |T_1|.$$

The above theorem, first proved by A.P. Morse [1947] by some more complicated means, still admits more abstract generalizations, exploiting the same idea. For these one can see Alfsen [1965].

(2) The theorem of Vitali with respect to the Lebesgue measure for regular sets.

Let P be a collection of measurable sets. We shall say that P is regular when there is a constant $c < \infty$ (regularity coefficient) such that for each $M \in P$ there exists a cubic interval Q verifying $Q \supset M$ and $|Q| \leq c \, |M|$. The following theorem concerning regular coverings is interesting.

Let A be a set in R^n. For each $x \in A$ a sequence $\{K_k(x)\}_{k=1}^{\infty}$ of compact sets such that $x \in K_k(x)$ and $\delta(K_k(x)) \to 0$ is given. Assume further that each sequence $\{K_k(x)\}_{k=1}^{\infty}$ is regular in the above sense with a regularity coefficient $c(x)$ that may depend on x. Then one can choose from $V = (K_k(x))_{x\in A,\ k=1,2,\ldots}$ a disjoint sequence $\{S_k\}$ such that $|A - \bigcup S_k| = 0$.

If the regularity coefficient satisfies $c(x) \leq H < \infty$ a simple modification of the proof of Theorem 3.1. gives this result. If $\sup \{c(x) : x \in A\} = \infty$, we first consider

$$A_1 = \{x \in A : c(x) \leq 1\}$$

and we apply the theorem to A_1, obtaining from V $\{S_k^1\}$ disjoint such that

$$|A_1 - \bigcup_k S_k^1| = 0.$$

We keep a finite number of sets of $\{S_k^1\}$, call them $\{S_k\}_1^{h_1}$, such that

$$|A_1 - \bigcup_{k=1}^{h_1} S_k|_e < \frac{1}{2}$$

Consider now

$$A_2 = \{x \in A - \bigcup_{k=1}^{h_1} S_k : c(x) \leq 2\}$$

and do the same, taking now only sets of V that are disjoint with $\bigcup_{k=1}^{h_1} S_k$ and obtaining $\{S_k\}_{h_1+1}^{h_2}$ such that

$$|A_2 - \bigcup_{k=h_1+1}^{h_2} S_k| < \frac{1}{2^2}$$

And so on. We obtain $\{S_k\}$ so that $|A - \bigcup S_k| = 0$.

The condition that the sets $K_k(x)$ are compact can be relaxed. It is sufficient that the boundary $\partial K_k(x)$ of each one of them have measure zero. In Remark (8) we shall have occasion to see an illuminating example related to this situation.

(3) The theorem of Vitali with respect to a general measure for other types of sets.

The structure of the proof of Theorem 3.2. is mainly based on the use of the theorem of Besicovitch. It is obvious that one can replace cubic intervals by any other collection of sets that satisfy a theorem of the type of Besicovitch, for example families such as those appearing in the remarks of Section 1.

(4) The cubes in Theorem 3.2. must be more or less centered at the corresponding points for $n \geq 2$.

In R^2 consider the measure μ that assigns to each Borel set $P \subset R^2$ the unidimensional Lebesgue measure of the intersection of P with the segment L joining the points (0,1) and (1,0). For each $x \in L$ let $\{Q_k(x)\}$ be the sequence of closed cubic intervals of diameter $\frac{1}{k}$ having x as lower left vertex. Then, for each sequence $\{S_k\}$ taken from $(Q_k(x))_{x\in L,\ k=1,2,\ldots}$ one has

$$\mu(L - \bigcup S_k) = \sqrt{2} > 0.$$

This example appears in Besicovitch [1945]. In R^1 however the situation is different, as the next remark shows. This different behavior is based on the total order of R^1.

(5) A general form of the Vitali theorem in R^1.

Let μ be a measure defined on the intervals of R^1 and μ^* the exterior measure associated to μ. Let $A \subset R^1$ be such that $\mu^*(A) < \infty$. For each $x \in A$ a sequence $\{I_k(x)\}$ is given of (nondegenerate) closed intervals containing x and such that $\delta(I_k(x)) \to 0$. Then one can select from $V = (I_k(x))_{x\in A,\ k=1,2,\ldots}$ a sequence $\{S_k\}$ of disjoint intervals such that

$$\mu^*(A - \bigcup S_k) = 0.$$

The proof of this theorem can be carried out by means of a principle analogous to that of the theorem of Besicovitch as it is done in Guzmán [1975]*. But it is still easier to obtain it by following the same line of the proof of Theorem 3.1. The reader is invited to do it as an exercise. Some results in the same direction are due to Iseki [1960], Jeffery [1934], [1958] and to Ellis and Jeffery [1967].

(6) An application. The Lebesgue density theorem.

With the theorem of Vitali one can obtain a quick proof of the density theorem of Lebesgue.

Let M be a measurable set of R^n. Then for almost all $x \in R^n$ one has, for each sequence $\{Q_k(x)\}$ of cubic intervals centered at x and contracting to x,

$$\lim_{k\to\infty} \frac{|Q_k(x) \cap M|}{|Q_k(x)|} = \chi_M(x)$$

We shall first prove that for any bounded measurable set $B \subset R^n$ and for any $a > 0$ the set

$$A_a = \{x \in B - M : \exists \{Q_k(x)\},\ Q_k(x) \to x,\ \frac{|Q_k(x)| \cap M|}{|Q_k(x)|} > a\}$$

is of measure zero. To do this, let $\varepsilon > 0$ and take an open set G containing A_a and such that $|G \cap M| \leq \varepsilon$. We apply to A_a the theorem of Vitali with cubes $\{Q_k(x)\}$ that appear in the definition of A_a and that are contained in G. Then one has $\{S_k\}$ verifying

$$|A_a - \bigcup_k S_k| = 0,\ S_k \subset G,\ \frac{|S_k \cap M|}{|S_k|} > a,\ S_k \cap S_j = \phi$$

if $k \neq j$.

Hence

$$|A_a|_e \leqslant \sum_k |S_k| \leqslant \frac{1}{a}\sum_k |S_k \cap M| \leqslant \frac{1}{a}\,|G \cap M| \leqslant \frac{\varepsilon}{a}$$

Since ε is arbitrary small, $|A_a| = 0$.

This proves that for almost every $x \notin M$ one has, for each sequence $\{Q_k(x)\}$ centered at x and contracting to x,

$$\lim \frac{|Q_k(x) \cap M|}{|Q_k(x)|} = 0 \text{ as } k \to \infty$$

Applying this to the complement M' of M we have for almost every $x \in M$, and each $\{Q_k(x)\}$

$$\lim \frac{|Q_k(x) \cap M'|}{|Q_k(x)|} = 0 \text{ as } k \to \infty$$

Since

$$\frac{|Q_k(x) \cap M'|}{|Q_k(x)|} = 1 - \frac{|Q_k(x) \cap M|}{|Q_k(x)|}$$

one obtains, at almost each $x \in M$, and each $Q_k(x)$

$$\lim \frac{|Q_k(x) \cap M|}{|Q_k(x)|} = 1 \quad \text{as} \quad k \to \infty$$

This concludes the proof.

(7) A more general application. Differentiation of set functions in R^n.

The following result includes many especial cases extremely useful, in particular part of the previous density theorem of Remark (6).

Let μ be a set function defined on finite unions of closed cubic intervals of R^n. Assume that μ is nonnegative, monotone, finitely additive and finite on each cube. Then at almost every (in the Lebesgue sense) point $x \in R^n$ one has, for each sequence $\{Q_k(x)\}$ of closed cubic intervals centered at x and contracting to x, that the limit

$$\lim_{k\to\infty} \frac{\mu(Q_k(x))}{|Q_k(x)|} = D(\mu,x)$$

exists, is finite and is independent of the sequence $\{Q_k(x)\}$. In order to prove this result, define first

$$A_\infty = \{x \in R^n : \exists\ \{Q_k(x)\},\ Q_k(x) \to x,\ \lim_{k\to\infty} \frac{\mu(Q_k(x))}{|Q_k(x)|} = \infty\}$$

We try to show that $|A_\infty| = 0$. Let us take an arbitrary closed cubic interval Q and a constant $M > 0$. For each $x \in A_\infty \cap \mathring{Q}$ we have $\{Q_k(x)\}$ contracting to x such that

$$\mu(Q_k(x)) > M\ |Q_k(x)|$$

and $Q_k(x) \subset \mathring{Q}$. We apply Theorem 3.1. obtaining a disjoint sequence $\{S_k\}$ from such cubes so that

$$|(A_\infty \cap \mathring{Q}) - \bigcup S_k| = 0$$

So we get

$$|A_\infty \cap \mathring{Q}|_e \leq \sum_k |S_k| < \frac{1}{M} \sum_k \mu(S_k) \leq \frac{1}{M}\ \mu(Q)$$

Since M is arbitrary and $\mu(Q) < \infty$ we obtain $|A_\infty \cap \mathring{Q}| = 0$ and so $|A_\infty| = 0$.

Let us now take an arbitrary closed cubic interval Q and define, for $r > s > 0$,

$$A_{rs} = \{x \in \mathring{Q} : \exists Q_k(x) \to x, \exists\ Q_k^*(x) \to x,\ Q_k(x) \subset \mathring{Q},\ Q_k^*(x) \subset \mathring{Q},$$

$$\frac{\mu(Q_k(x))}{|Q_k(x)|} > r > s > \frac{\mu(Q_k^*(x))}{|Q_k^*(x)|}\}$$

We try to prove that $|A_{rs}| = 0$. Let $\varepsilon > 0$ arbitrary and G an open set containing A_{rs} such that $|G| \leq |A_{rs}|_e + \varepsilon$. We apply Vitali's theorem to A_{rs} with the cubes $Q_k^*(x)$, taking only those contained in G and we obtain a disjoint sequence $\{S_k^*\}$ of such cubes so that $|A_{rs} - \bigcup_k S_k^*| = 0$. We clearly have

$$|G| \geq \sum_k |S_k^*| \geq \frac{1}{s} \sum_k \mu(S_k^*)$$

Observe that, if

$$C = A_{rs} \bigcap (\bigcup_k \overset{\circ}{S_k^*}),$$

we have $|C|_e = |A_{rs}|_e$. For each $x \in C$ there is an S_j^* such that $x \in \overset{\circ}{S_j^*}$ and so, since $x \in A_{rs}$, there is also a sequence $Q_k(x) \to x$ such that $Q_k(x) \subset \overset{\circ}{S_j^*}$ and

$$\frac{\mu(Q_k(x))}{|Q_k(x)|} > r.$$

We now apply the Vitali theorem again to C with these cubes and so obtain a disjoint sequence $\{S_k\}$ such that each S_k is in some S_j^* and $|C - \bigcup S_k| = 0$. Thus we have, taking into account the above inequalities

$$|A_{rs}|_e = |C|_e = |C \bigcap (\bigcup S_k)|_e \leq \sum_k |S_k| \leq \frac{1}{r} \sum_k \mu(S_k) \leq \frac{1}{r} \sum_k \mu(S_k^*) \leq$$

$$\leq \frac{s}{r} |G| \leq \frac{s}{r} (|A_{rs}|_e + \varepsilon).$$

Thus

$$|A_{rs}|_e \leq \varepsilon \frac{s}{r-s}.$$

Since ε is arbitrarily small we obtain $|A_{rs}| = 0$. With this one easily concludes the proof of the theorem.

We obtain in particular the following. If P is an arbitrary set of R^n and we set for each closed cubic interval $\mu_1(Q) = |Q \cap P|_e$, we obtain that the limit

$$\lim_{k\to\infty} \frac{|Q_k(x) \cap P|_e}{|Q_k(x)|} = d_e(x)$$

exists and is finite at almost every x.

We can also take $\mu_2(Q) = |Q \cap P|_i$, i.e. the interior measure of $Q \cap P$, and so

$$\lim_{k\to\infty} \frac{|Q_k(x) \cap P|_i}{|Q_k(x)|} = d_i(x)$$

exists and is finite at almost every x.

If $\mu(Q) = \int_Q f$ with f nonnegative and locally integrable, we get that

$$\lim_{k\to\infty} \frac{1}{|Q_k(x)|} \int_{Q_k(x)} f = h(x)$$

exists, is finite and independent of the particular $\{Q_k(x)\}$ at almost every $x \in R^n$.

(8) A collection of sets that does not satisfy the Vitali property.

Further on we shall see that if in theorem 3.1. we substitute the closed cubic intervals $\{Q_k(x)\}$ by closed intervals $\{I_k(x)\}$ centered at x and contracting to x, then we cannot select in general from $(I_k(x))_{x\in A,\ k=1,2,\dots}$ a disjoint sequence covering almost all of A. The proof of this fact will be presented in Chapter IV. Here we present an easy example of Hayes [1952] of a family of sets which has very good differentiation properties but does not satisfy the Vitali property.

Let A be the closed unit cube of R^2 and for each $x \in A$ let $\{S_k(x)\}$ be the following sequence of sets: For k = 1, 2, ..., $S_k(x)$ is the closed cubic interval of center x and side-length $\frac{1}{8k}$ together with all points with rational coordinates contained in the closed cubic interval of center x and side-length $\frac{1}{4k}$.

It is easy to prove that for any disjoint sequence $\{T_k\}$ taken from $(S_k(x))_{x \in A,\ k=1,2,\dots}$ due to the cloud of points surrounding the sets $Q_k(x)$ one has $|A - T_k| > 0$.

However, as we shall see later on, this family of sets $(S_k(x))_{x \in A,\ k=1,2,\dots}$ has very good differentiation properties

(9) Lebesgue's differentiation theorem as a consequence of the theorem of Vitali.

Let $f \in L^1(R^n)$. We want to prove that the set of points $x \in R^n$ for which there is a sequence $\{Q_k(x)\}$ of closed cubic intervals centered at x and contracting to x such that

$$\frac{1}{|Q_k(x)|} \int_{Q_k(x)} f$$

does not converge to f(x) is a null set. We define, for $\alpha > 0$, $H > 0$, the set

$$A = \{x \in R^n : |x| \leq H, \exists \{Q_k(x)\},\ Q_k(x) \longrightarrow x,$$

$$\limsup_{k\to\infty} \left| \frac{1}{|Q_k(x)|} \int_{Q_k(x)} f - f(x) \right| > \alpha\}$$

It will be enough to prove $|A| = 0$. Let $\varepsilon > 0$ and set $f = g + h$ with g continuous and $||h||_1 < \varepsilon$. Then

$$A = \{x \in R^n : |x| \leq H, \exists \{Q_k(x)\},\ Q_k(x) \longrightarrow x,$$

$$\limsup_{k\to\infty} \left| \frac{1}{|Q_k(x)|} \int_{Q_k(x)} h - h(x) \right| > \alpha\}$$

since, for g continuous, clearly we have

$$\frac{1}{|Q_k(x)|}\int_{Q_k(x)} g \longrightarrow g(x)$$

everywhere. We can write

$$A \subset \{x \in R^n \; |x| \leq H, \exists \{Q_k(x)\},\; Q_k(x) \to x,\; \frac{1}{|Q_k(x)|}\int_{Q_k(x)} |h| > \frac{\alpha}{2}\} \cup$$

$$\cup \{x \in R^n : |h(x)| > \frac{\alpha}{2}\} = A_1 \cup A_2.$$

We easily get

$$|A_2| = \int_{A_2} dx \leq \int \frac{|h(x)|}{\alpha/2}\, dx \leq \frac{2\varepsilon}{\alpha}$$

Applying Vitali's theorem to A_1 we get a disjoint sequence $\{S_k\} \subset (Q_k(x))_{x\in A, k\geq 1}$ such that $|A_1 - \cup S_k| = 0$ and

$$\frac{1}{|S_k|}\int_{S_k} |h| > \frac{\alpha}{2}$$

So we obtain

$$|A_1|_e \leq \sum |S_k| \leq \frac{2}{\alpha}\sum \int_{S_k} |h| \leq \frac{2\varepsilon}{\alpha}.$$

Therefore $|A|_e \leq \frac{4\varepsilon}{\alpha}$. Since $\varepsilon > 0$ is arbitrary, $|A| = 0$.

CHAPTER II

THE HARDY-LITTLEWOOD MAXIMAL OPERATOR

For a function $f \in L_{loc}(R)$ Hardy and Littlewood [1930] introduced a new function Mf that plays a fundamental role in real variable theory. Such a function can be defined on each $x \in R$ by

$$Mf(x) = \sup_{h>0} \frac{1}{2h} \int_{x-h}^{x+h} |f(s)|ds.$$

The function Mf is measurable, since the set $\{Mf > \lambda\}$ is open for each $\lambda > 0$, and the operator M

$$M : f \in L_{loc}(R) \to Mf \in \mathcal{M}(R),$$

where $\mathcal{M}(R)$ denotes the space of all real valued measurable functions defined on R, is called the Hardy-Littlewood maximal operator. We shall see several generalizations of this operator, whose behavior is very informative in particular in differentiation theory. The most natural among these generalizations consists in defining, for $f \in L_{loc}(R^n)$ at each $x \in R^n$

$$Mf(x) = \sup_{r>0} \frac{1}{|Q(x,r)|} \int_{Q(x,r)} |f(y)|dy$$

where $Q(x,r)$ stands for the open cubic interval of center x and side length 2r. The operator M defined in this way from $f \in L_{loc}(R^n)$ to $\mathcal{M}(R^n)$ satisfies the following obvious properties:

i) For each $f \in L_{loc}(R^n)$ and $x \in R^n$, $Mf(x) \geq 0$.

ii) For each $f_1, f_2 \in L_{loc}(R^n)$,

$$M(f_1+f_2)(x) \leq Mf_1(x) + Mf_2(x).$$

iii) For each $f \in L_{loc}(R^n)$, $\lambda \in R$, $x \in R^n$, $M(\lambda f) = |\lambda|\, Mf(x)$.

We shall refer to these properties saying that M is positive (property i)), subadditive (property ii)) and positively homogeneous (property iii)).

In the following sections of this chapter we shall present several interesting properties of the Hardy-Littlewood maximal operator and of some natural generalizations of it that are useful in differentiation theory.

1. WEAK TYPE (1,1) OF THE MAXIMAL OPERATOR.

The maximal operator defined in the introduction obviously satisfies

$$||Mf||_\infty \leq ||f||_\infty.$$

We shall now see that it satisfies another important inequality.

An operator $T : L_{loc}(R^n) \to \mathfrak{M}(R^n)$ which is subadditive (i.e. such that for each $f_1, f_2 \in L_{loc}(R^n)$ and for almost every $x \in R^n$ we have

$$|T(f_1+f_2)(x)| \leq |Tf_1(x)| + |Tf_2(x)|)$$

is said to be of weak type (p,p) where $1 \leq p < \infty$ when the following inequality holds for each $\lambda > 0$

$$|\{x : |Tf(x)| > \lambda\}| \leq \left(\frac{c||f||_p}{\lambda}\right)^p$$

c being a constant independent of f and λ.

The operator T is said to be of strong type (p,p), where now $1 \leq p \leq \infty$ if for every $f \in L_{loc}(R^n)$,

$$||Tf||_p \leq c||f||_p,$$

with c independent of f. Observe that if T is of strong type (p,p) for a p with $1 \leq p < \infty$, then T is of weak type (p,p), since we have, for any $f \in L_{loc}(R^n)$ and $\lambda > 0$,

$$|A_\lambda| = |\{x : |Tf(x)|>\lambda\}| = \int \chi_{A_\lambda}(x)dx \leq \int \frac{|Tf(x)|^p}{\lambda^p} dx \leq (\frac{c||f||_p}{\lambda})^p.$$

In the remark (2) of this section we shall indicate the importance of this notions. Here we show how the theorem of Besicovitch of I.1 leads inmediately to the weak type (1,1) of the maximal operator we have introduced in Section 1.

THEOREM 2.1. <u>The Hardy-Littlewood maximal operator introduced above is of weak type</u> (1,1).

<u>Proof</u>. Let $f \in L_{loc}(R^n)$ and $\lambda > 0$. We consider

$$A = \{x \in R^n : Mf(x) > \lambda\}.$$

Let K be any bounded measurable set of R^n. For each $x \in A \cap K$ we choose an open cubic interval Q(x) centered at x such that

$$\frac{1}{|Q(x)|} \int_{Q(x)} |f(y)|dy > \lambda$$

We can apply the theorem of Besicovitch (cf. also Remark (2) of I.1.) obtaining $\{Q_k\}$ from $(Q(x))_{x \in A \cap K}$ such that $A \cap K \subset \bigcup Q_k$ and, if χ_k is the characteristic function of Q_k, we have $\sum \chi_k \leq \theta_n$, where θ_n is constant of the theorem. Hence

$$|A \cap K| \leq |\bigcup_k Q_k| \leq \sum_k |Q_k| \leq \frac{1}{\lambda} \sum_k \int_{Q_k} |f| = \frac{1}{\lambda} \int_{\bigcup Q_k} |f| \sum_k \chi_k \leq$$

$$\leq \frac{\theta_n}{\lambda} \int_{\bigcup Q_k} |f| = \frac{\theta_n}{\lambda} ||f||_1.$$

Since this estimate is independent of K, we obtain

$$|A| \leq \frac{\Theta_n}{\lambda} ||f||_1$$

and so the theorem is proved.

REMARKS.

(1) A refinement of the theorem.

One can refine the inequality we have obtained in order to get

$$|\{Mf > \lambda\}| \leq \frac{2\Theta_n}{\lambda} \int_{|f|>\frac{\lambda}{2}} |f|$$

In fact, we define

$$f_*(x) = \begin{cases} 0 & \text{if } |f(x)| > \frac{\lambda}{2} \\ f(x) & \text{if } |f(x)| \leq \frac{\lambda}{2} \end{cases}$$

and $f^*(x)$ such that $f(x) = f^*(x) + f_*(x)$.

It is clear, since $Mf(x) \leq Mf^*(x) + Mf_*(x)$, that

$$\{Mf > \lambda\} \subset \{Mf^* > \tfrac{\lambda}{2}\} \cup \{Mf_* > \tfrac{\lambda}{2}\}$$

However, $\{Mf_* > \frac{\lambda}{2}\} = \phi$ and so f

$$|\{Mf > \lambda\}| \leq |\{Mf^* > \tfrac{\lambda}{2}\}| \leq \frac{2\Theta_n}{\lambda} \int |f^*| = \frac{2\Theta_n}{\lambda} \int_{|f|>\frac{\lambda}{2}} |f|$$

(2) Strong type (p,p), $1 < p \leq \infty$, of the maximal operator.

Since M is of strong type (∞,∞) and of weak type $(1,1)$ the interpolation theorem of Marcinkiewicz (cf. Zygmund [1959], vol.II p.111) allows us to conclude that M is of strong type (p,p) for each p such that $1 < p < \infty$.

That the maximal operator is not of strong type (1,1) is clear from the following example. In R let f be the characteristic function of the unit interval $[0,1]$. Then $Mf(x) = \frac{1}{2x}$ for $x \geq 1$ and so $\int |Mf| = \infty$.

(3) An application. The Lebesgue differentiation theorem.

The theorem we have obtained leads us to an easy proof of the following result:

Let $f \in L(R^n)$. *Then for almost every* $x \in R^n$, *we have*

$$\lim_{r \to 0} \frac{1}{|Q(x,r)|} \int_{Q(x,r)} f = f(x)$$

where $Q(x,r)$ *denotes the open cubic interval centered at* x *and side length* $2r$.

For the proof of this fact we shall try to show that for each $\lambda > 0$ the set

$$P_\lambda = \{x \in R^n : \limsup_{r \to 0} |\frac{1}{|Q(x,r)|} \int_{Q(x,r)} f - f(x)| > \lambda\}$$

is of measure zero. This obviously implies the theorem.

We prove $|P_\lambda| = 0$ in the following way. Given $\varepsilon > 0$ we take a continuous function g such that the function $h = f - g$ has an L^1-norm $||h||_1 \leq \varepsilon$. For g we have clearly at each $x \in R^n$

$$\lim_{r \to 0} \frac{1}{|Q(x,r)|} \int_{Q(x,r)} g = g(x)$$

and so

$$P_\lambda = \{x \in R^n : \limsup_{r \to 0} |\frac{1}{|Q(x,r)|} \int_{Q(x,r)} h - h(x)| > \lambda\} \subset$$

$$\subset \{x \in R^n : \limsup_{r\to 0} |\frac{1}{|Q(x,r)|}\int_{Q(x,r)} h| > \frac{\lambda}{2}\} \cup \{x \in R^n : |h(x)| > \frac{\lambda}{2}\} =$$

$$= P_\lambda^1 \cup P_\lambda^2 .$$

Since

$$\limsup_{r\to 0} \frac{1}{|Q(x,r)|}\int_{Q(x,r)} |h| \leq Mh(x),$$

we have

$$|P_\lambda^1|_e \leq |\{x : Mh(x) > \frac{\lambda}{2}\}| \leq \frac{2\Theta_n}{\lambda} ||h||_1 \leq \frac{2\Theta_n \varepsilon}{\lambda}$$

Also

$$|P_\lambda^2| = |\{x : |h(x)| > \frac{\lambda}{2}\}| \leq \frac{2||h||_1}{\lambda} \leq \frac{2\varepsilon}{\lambda}$$

Since $\varepsilon > 0$ is arbitrary, $|P_\lambda| = 0$.

The theorem we have proved is of course valid if instead or assuming $f \in L(R^n)$ we just assume $f \in L_{loc}(R^n)$. This fact is due to the strictly local character of the relation

$$\lim_{r\to 0} \frac{1}{|Q(x,r)|}\int_{Q(x,r)} f = f(x)$$

we have proved, i.e. it depends only on the behavior of f in a neighborhood of x. If $f \in L_{loc}(R^n)$ and χ_H is the characteristic function of the open Euclidean ball with center 0 and radius $H > 0$, then $h = f\chi_H \in L(R^n)$. We apply then the theorem to h and this gives us the result we are seeking for f at almost all points inside B(0,H). Since H is arbitrary we obtain it for f at almost every point.

This observation permits us to prove a result a little finer.

Let $f \in L_{loc}(R^n)$. Then, for almost every $x \in R^n$, we have

$$\lim_{r\to 0} \frac{1}{|Q(x,r)|} \int_{Q(x,r)} |f(y) - f(x)|dy = 0.$$

To prove this, let $\{q_k\}_{k=1}^{\infty}$ be an enumeration of all rational numbers. For each k we consider the function $|f(\cdot) - q_k|$, which is obviously in $L_{loc}(R^n)$. Then if for each k we exclude a nullset E_k, we have, for each $x \notin E_k$

$$\lim_{r\to 0} \frac{1}{|Q(x,r)|} \int_{Q(x,r)} |f(y) - q_k|dy = |f(x) - q_k|.$$

Let $E = \bigcup_{k=1}^{\infty} E_k$. Obviously $|E| = 0$ and if $z \notin E$, then, for each K

$$\lim_{r\to 0} \frac{1}{|Q(z,r)|} \int_{Q(z,r)} |f(y) - q_k|dy = |f(z) - q_k|$$

Let

$$A = \{x \in R^n : |f(x)| = \infty\}.$$

Clearly $|A| = 0$. Take now a fixed $v \notin A \bigcup E$ and $\varepsilon > 0$. Let q_k be a rational number such that $|f(v) - q_k| \leq \frac{\varepsilon}{2}$.

Now we can write

$$\limsup_{r\to 0} \frac{1}{|Q(v,r)|} \int_{Q(v,r)} |f(y) - f(v)|\, dy \leq$$

$$\leq \limsup_{r\to 0} \frac{1}{|Q(v,r)|} \int_{Q(v,r)} |f(y) - q_k|\, dy +$$

$$+ \limsup_{r\to 0} \frac{1}{|Q(v,r)|} \int_{Q(v,r)} |q_k - f(v)|dy = 2|f(v) - q_k| \leq \varepsilon.$$

Since ε is arbitrary we get the result.

2. DIFFERENTIATION BASES AND THE MAXIMAL OPERATOR ASSOCIATED TO THEM.

Here we present a generalization of the Hardy-Littlewood maximal operator that will be very useful in the problems on differentiation we shall consider.

For each $x \in R^n$ let $\mathcal{B}(x)$ be a collection of bounded measurable sets with positive measure containing x and such that there is at least a sequence $\{R_k\} \subset \mathcal{B}(x)$ with $\delta(R_k) \to 0$. The whole collection $\mathcal{B} = \bigcup_{x \in R^n} \mathcal{B}(x)$ will be called a differentiation basis.

For example if $\mathcal{B}_1(x)$ is the collection of all open bounded cubic intervals containing x we obtain a basis $\mathcal{B}_1$. Analogously $\mathcal{B}_2$ will denote the basis such that $\mathcal{B}_2(x)$ is the collection of all open bounded intervals containing x and $\mathcal{B}_3$ the basis such that $\mathcal{B}_3(x)$ is the collection of all open bounded rectangular parallelepipeds containing x.

We denote by $\mathcal{B}_1^*$ the basis such that $\mathcal{B}_1^*(x)$ is the collection of all open bounded cubic intervals centered at x. Similarly $\mathcal{B}_2^*$ and $\mathcal{B}_3^*$. Also we shall denote by $\mathcal{S}$ the basis such that $\mathcal{S}(x)$ is the family of all open Euclidean balls containing x and $\mathcal{S}^*$ will stand for the basis such that $\mathcal{S}^*(x)$ is the family of all open Euclidean balls centered at x.

A basis $\mathcal{B}$ of open sets will be called a Busemann-Feller basis (B-F basis) whenever for each $R \in \mathcal{B}$ and $x \in R$ we have $R \in \mathcal{B}(x)$. The bases $\mathcal{B}_1, \mathcal{B}_2, \mathcal{B}_3, \mathcal{S}$ are B-F bases, but the corresponding bases $\mathcal{B}_1^*, \mathcal{B}_2^*, \mathcal{B}_3^*, \mathcal{S}^*$ are not. Certain measurability problems arising in differentiation theory are very easy to handle for B - F bases and can be difficult for other bases.

For a differentiation basis $\mathcal{B}$ and a function $f \in L_{loc}(R^n)$ we define for each $x \in R^n$

$$Mf(x) = \sup \{\frac{1}{|R|} \int_R |f| \; : \; R \in \mathcal{B}(x)\}$$

If $\mathcal{B}$ is a B - F basis it is clear that the set $\{Mf > \lambda\}$ for any $\lambda > 0$ is open and so M is an operator from $L_{loc}(R^n)$ to $\mathcal{M}$. The measurability of Mf (also trivial for $\mathcal{B}_1^*, \mathcal{B}_2^*, \mathcal{B}_3^*, \mathcal{S}^*$) can be a difficult problem in some other instances. In any case M is an operator sending L_{loc} into the space of all functions defined on R^n with values in $[0,\infty]$. We shall call M the maximal operator associated to $\mathcal{B}$.

It is clear that M satisfies properties i), ii), iii) of the introduction of this chapter, i.e. it is positive, subadditive and positively homogeneous.

Observing the arguments of the preceding section we can see that all that has been said about the Hardy-Littlewood maximal operator there has been based on a property of the Besicovitch type. Let us say that a basis $\mathcal{B}$ satisfies the Besicovitch property whenever: Given a bounded set A of R^n and for each $x \in A$ a set $R(x) \in \mathcal{B}(x)$, one can choose from $(R(x))_{x\in A}$ a sequence $\{R_k\}$ so that

i) $A \subset \bigcup R_k$.

ii) Each $z \in R^n$ is at most in θ (a number depending only on $\mathcal{B}$) of the sets R_k, i.e. $\sum \chi_{R_k} \leq \theta$.

iii) The sequence $\{R_k\}$ can be distributed into ξ (a number depending only on $\mathcal{B}$) disjoint sequences.

Without any substantial change in the proof with respect to those given in the preceding section and their remarks we get the following result.

THEOREM 2.1. Let $\mathcal{B}$ be a basis satisfying the Besicovitch property. Then its maximal operator satisfies, for each $f \in L_{loc}(R^n)$ and each $\lambda > 0$.

$$|\{Mf > \lambda\}|_e \leq \frac{\theta}{\lambda} ||f||_1$$

and also for almost each $x \in R^n$

$$\lim \frac{1}{|R|} \int_R |f(y) - f(x)| \, dy = 0$$

for $R \in \mathcal{B}(x)$ and $\delta(R) \to 0$.

REMARKS.

(1) The conclusion of Theorem 2.1. is valid for $\mathcal{B}_1, \mathcal{S}$.

The bases $\mathcal{B}_1, \mathcal{S}$ do not satisfy the Besicovitch property, as can be concluded from the example of Remark (4) of Section I.3. However, if M_1 is the maximal operator of $\mathcal{B}_1$ and M_1^* that of $\mathcal{B}_1^*$ one has, for each $f \in L_{loc}(R^n)$ and each $x \in R^n$

$$M_1^*f(x) \leq M_1f(x) \leq 2^n M_1^*f(x)$$

In fact, if $Q \in \mathcal{B}(x)$ there exists $Q^* \in \mathcal{B}_1^*(x)$, with $Q^* \supset Q$ such that $|Q^*| \leq 2^n|Q|$ (Take Q^* the minimum open cubic interval centered at x and containing Q). Then we can write

$$\frac{1}{|Q|}\int_Q |f| = \frac{2^n}{2^n|Q|}\int_Q |f| \leq \frac{2^n}{|Q^*|}\int_{Q^*} |f|$$

and so we obtain $M_1f(x) \leq 2^n M_1^* f(x)$. The other inequality is trivial. This permits us to infer that the conclusion of Theorem 2.1. is also valid for $\mathcal{B}_1$. The same is true for $\mathcal{S}$.

(2) A natural question on $\mathcal{B}_2, \mathcal{B}_3$.

A question that arises naturally when we consider this generalization of the maximal operator consists in finding out whether $\mathcal{B}_2, \mathcal{B}_3, \mathcal{B}_2^*, \mathcal{B}_3^*$ and their maximal operators satisfy properties similar to those of $\mathcal{B}_1, \mathcal{B}_1^*$. The following section handles the case of $\mathcal{B}_2$.

3. THE MAXIMAL OPERATOR ASSOCIATED TO A PRODUCT OF DIFFERENTIATION BASES.

The main purpose of this section is to establish for $\mathcal{B}_2$ and its maximal operator M_2 a weak type inequality of the type that has been proved for $\mathcal{B}_1$ in the preceding section. The basic idea of the method can be shown in a simple way for R^2. In the remarks at the end of this section we indicate the modifications one has to introduce in order to obtain a similar theorem for R^n with $n > 2$ and for some other more general differentiation bases. The material presented in this section belongs to Guzmán [1974].

THEOREM 3.1. *Let M_2 be the maximal operator associated to the interval basis $\mathcal{B}_2$ in R^2. Then, for each $f \in L_{loc}(R^2)$ and each $\lambda > 0$ we have*

$$|\{M_2 f > \lambda\}| \leq c \int \frac{|f|}{\lambda} (1 + \log^+ |f|)$$

where c is a positive constant, independent of f and λ, and $\log^+ a = 0$ if $0 \leq a \leq 1$ and $\log^+ a = \log a$ if $a > 1$.

Proof. We present here a proof of the theorem disregarding the easy, but tedious, measurability problems that arise in it. In the remark (1) following the theorem we give some indications and a reference to handle them.

In the proof we shall indicate by $|P|_1$ and $|Q|$ the Lebesgue measure of the measurable set $P \subset R^1$ and $Q \subset R^2$ respectively. For the sake of clarity we shall denote by Greek letters (ξ^1,ξ^2), (η^1,η^2), ... the dummy variable coordinates which appear in the integrals and definitions.

Let $f \geq 0$, $f \in L_{loc}(R^2)$. For $(x^1, x^2) \in R^2$ we define

$$T_1 f(x^1,x^2) = \sup \{\frac{1}{|J|_1} \int_J f(\xi^1,x^2) d\xi^1 : J \text{ interval of } R^1,\ x^1 \in J\}$$

For $\lambda > 0$ we consider the set

$$A = \{(\eta^1,\eta^2) \in R^2 : T_1 f(\eta^1,\eta^2) > \frac{\lambda}{2}\}$$

and again we define, for $(x^1,x^2) \in R^2$

$$T_2f(x^1,x^2) = \sup \{\frac{1}{|H|_1}\int_H \chi_A(x^1,\eta^2)T_1f(x^1,\eta^2)d\eta^2 : H \text{ interval of } R^1,\ x^2 \in H\}.$$

We shall first prove the relation

$$B = \{(\xi^1,\xi^2)\in R^2 : M_2f(\xi^1,\xi^2)>\lambda\} \subset \{(\xi^1,\xi^2)\in R^2 : T_2f(\xi^1,\xi^2)>\frac{\lambda}{2}\} = C.$$

Take a fixed point (x^1,x^2) of B. We wish to prove that $(x^1,x^2)\in C$. Since $(x^1,x^2)\in B$, there is an interval $I = J \times H$ of R^2 such that $x^1 \in J$, $x^2 \in H$ and

$$\frac{1}{|I|}\int_I f > \lambda$$

We now partition the interval I into two sets C_1, C_2, each one being a union of segments of the size of J parallel to the axis Ox^1 in the following way. Let $J \times \{y^2\}$ be one of such segments. If for each point $(z^1,y^2) \in J \times \{y^2\}$ we have $T_1f(z^1,y^2) > \frac{\lambda}{2}$ we set $J \times \{y^2\} \subset C_1$. Otherwise, i.e. if there is some point $(z^1,y^2) \in J \times \{y^2\}$ such that $T_1f(z^1,y^2) \leq \frac{\lambda}{2}$ we set $J \times \{y^2\} \subset C_2$. Observe that $J \times \{y^2\} \subset C_2$ implies in particular that

$$\frac{1}{|J|_1}\int_J f(\xi^1,y^2)\,d\xi^1 \leq \frac{\lambda}{2}$$

and so, integrating this inequality over the set G of all ξ^2 in H such that $J \times \{\xi^2\} \subset C_2$, we get

$$\int_{C_2} f = \int_G\int_J f(\xi^1,\xi^2)d\xi^1 d\xi^2 \leq \int_G \frac{\lambda}{2}\,|J|_1 d\xi^2 = \frac{\lambda}{2}\,|C_2| \leq \frac{\lambda}{2}\,|I|$$

Since

$$\int_{C_1} f + \int_{C_2} f = \int_I f > \lambda\,|I|, \text{we get} \int_{C_1} f > \frac{\lambda}{2}\,|I|.$$

We can also write, by virtue of the definition of T_2 and of T_1,

$$T_2f(x^1,x^2) \geq \frac{1}{|H|_1}\int_H \chi_A(x^1,\eta^2)T_1f(x^1,\eta^2)T_1f(x^1,\eta^2)d\eta^2 \geq$$

$$\geq \frac{1}{|H|_1}\int_H \chi_A(x^1,\eta^2)\ \frac{1}{|J|_1}\int_J f(\eta^1,\eta^2)d\eta^1\ d\eta^2 \geq$$

$$\geq \frac{1}{|I|}\int_H \chi_A(x^1,\eta^2)\int_J f(\eta^1,\eta^2)\ d\eta^1\ d\eta^2$$

By the definition of C_1 and A, if $(\eta^1,\eta^2) \in C_1$ then $(x^1,\eta^2) \in A$ and so the last member of the above chain of inequalities is

$$\geq \frac{1}{|I|}\int_{C_1} f(\eta^1,\eta^2)\ d\eta^1\ d\eta^2 > \frac{\lambda}{2}$$

This concludes the proof that $B \subset C$. We now prove that C satisfies the inequality we are looking for.

We can assume $f \in L(1 + \log^+ L)$, since otherwise there is nothing to prove. In the following argument c will be an absolute constant not always the same in each ocurrence, independent in particular of f and λ.

By virtue of the weak type (1,1) for the unidimensional basis $\mathcal{B}_1$ of intervals, for almost each fixed $x^1 \in R$ we can write

$$|\{\xi^2 \in R : T_2f(x^1,\xi^2) > \frac{\lambda}{2}\}|_1 \leq \frac{c}{\lambda}\int_R \chi_A(x^1,\xi^2)T_1f(x^1,\xi^2)d\xi^2$$

Hence, if we integrate over all such $x^1 \in R$ and interchange the order of integration, we get

$$|C| = |\{(\xi^1,\xi^2) \in R^2 : T_2f(\xi^1,\xi^2) > \frac{\lambda}{2}\}| \leq$$

$$\leq c\int_R\int_R \frac{\chi_A(\xi^1,\xi^2)\ T_1f(\xi^1,\xi^2)}{\lambda/2}\ d\xi^1\ d\xi^2 =$$

$$= c\int_R \int_0^\infty |\{\xi^1 \in R : \frac{\chi_A(\xi^1,\xi^2)T_1f(\xi^1,\xi^2)}{\lambda/2} > \sigma\}|_1 d\sigma\, d\xi^2 =$$

$$= [c\int_R \int_0^1] + [c\int_R \int_1^\infty] = S_1 + S_2$$

If $(\xi^1,\xi^2) \in A$, then $T_1f(\xi^1,\xi^2) > \frac{\lambda}{2}$, and so if $0 < \sigma \leq 1$, we have

$$|\{\xi^1 \in R : \frac{\chi_A(\xi^1,\xi^2)T_1f(\xi^1,\xi^2)}{\lambda/2} > \sigma\}|_1 = |\{\xi^1 \in R : T_1f(\xi^1,\xi^2) > \frac{\lambda}{2}\}|_1$$

Hence, by the weak type (1,1) for $\mathcal{B}_1$ in R^1, we get

$$S_1 = c\int_R |\{\xi^1 \in R : T_1f(\xi^1,\xi^2) > \frac{\lambda}{2}\}|\, d\xi^2 \leq c\int_R \int_R \frac{f(\xi^1,\xi^2)}{\lambda}\, d\xi^1\, d\xi^2$$

In order to estimate S_2 we define for a fixed $\sigma > 0$

$$f_*(\xi^1, \xi^2; \sigma) = \begin{cases} f(\xi^1,\xi^2) & \text{if } f(\xi^1,\xi^2) \leq \frac{\lambda\sigma}{4} \\ 0 & \text{if } f(\xi^1,\xi^2) > \frac{\lambda\sigma}{4} \end{cases}$$

and $f^*(\xi^1,\xi^2;\sigma)$ such that

$$f(\xi^1,\xi^2) = f^*(\xi^1,\xi^2;\sigma) + f_*(\xi^1,\xi^2;\sigma)$$

For brevity let us write $f = f^*_\sigma + f^\sigma_*$. It is clear that

$$T_1f \leq T_1f^*_\sigma + T_1f^\sigma_*$$

and $T_1f^\sigma_* \leq \frac{\lambda\sigma}{4}$. Hence,

$$S_2 \leq c\int_R \int_1^\infty |\{\xi^1 \in R : T_1f^*_\sigma(\xi^1,\xi^2) > \frac{\lambda\sigma}{4}\}|_1\, d\sigma\, d\xi^2 \leq$$

$$\leq c\int_R \int_1^\infty \int_R \frac{f^*_\sigma(\xi^1,\xi^2)}{\lambda\sigma/4}\, d\xi^1 d\sigma d\xi^2 \leq$$

$$\leq c \int_R \int_R \left(\int_1^{\frac{4f(\xi^1,\xi^2)}{\lambda\sigma}} \frac{4f(\xi^1,\xi^2)}{\lambda} \, d\sigma \right) d\xi^1 \, d\xi^2 =$$

$$= c \int_R \int_R \frac{f(\xi^1,\xi^2)}{\lambda} \log^+ \frac{4f(\xi^1,\xi^2)}{\lambda} \, d\xi^1 \, d\xi^2$$

Adding up we get

$$|C| \leq c \int \frac{f}{\lambda} (1 + \log^+ \frac{f}{\lambda})$$

and this implies inequality of the theorem.

For $\mathcal{B}_2^*$ and M_2^* in R^2 one also obtains and inequality of the same type, since as we have seen in the remark (1) of the previous section for M_1 and M_1^*, we also have here

$$M_2^* f(x) \leq M_2 f(x) \leq 2^2 M_2^* f(x)$$

for each $f \in L_{loc}(R^2)$ and each $x \in R^2$.

REMARKS.

(1) The measurability problem in Theorem 3.1.

If, in the proof of the theorem, we first consider a function f of the type

$$f(x^1,x^2) = \sum_{k=1}^{N} a_k \chi_{A_k}(x^1,x^2)$$

Where $a_k > 0$ and A_k is the product of two open bounded intervals in R, then the measurability of the sets which appear in the proof is rather simple. One then proves the inequality: (a) for a linear combination with nonnegative coefficients of characteristic functions of bounded open disjoint sets; (b) for a nonnegative linear combination of characteristic functions of disjoint compact sets; (c) for a nonnegative function of L_0; (d) for a function in L_{loc}. For details we refer to Guzmán [1974].

(2) A generalization of Theorem 3.1.

In a more general context one can obtain the following result, proceeding in exactly the same way as in the proof of the theorem:

Let $\mathcal{R}_i$, i = 1, 2, be a differentiation basis in R^{n_i} and let H_i be the maximal operator associociented to $\mathcal{R}_i$. Assume that H_i satisfies the following weak type inequality for each $\lambda > 0$ and $f_i \in L_{loc}(R^{n_i})$

$$m_i(\{x^i \in R^{n_i} : H_i f_i(x^i) > \lambda\}) \leq \int \phi_i\left(\frac{|f_i(x^i)|}{\lambda}\right) dm_i(x^i)$$

where m_i means n_i-dimensional Lebesgue measure and ϕ_i is a strictly increasing continuous function from $[0,\infty]$ to $[0,\infty]$ with $\phi_i(0) = 0$. Let $\mathcal{R}$ be the basis in $R^n =$ $= R^{n_1} \times R^{n_2}$ product of the two bases $\mathcal{R}_1, \mathcal{R}_2$, i.e.

$$\{B^1 \times B^2 : B^1 \in \mathcal{R}_1,\ B^2 \in \mathcal{R}_2\}$$

and let H be its maximal operator. Then, for each $\lambda > 0$ and each $f \in L_{loc}(R^n)$ one has

$$m(\{(x^1,x^2) \in R^{n_1+n_2} : Hf(x^1,x^2) > \lambda\}) \leq$$

$$\leq \phi_2(1) \int_{R^{n_1+n_2}} \phi_1\left(\frac{2|f|}{\lambda}\right) dm + \int_{R^{n_1+n_2}} \left[\int_1^{4f} \phi_1\left(\frac{4|f|}{\lambda}\right) d\phi_2(\sigma)\right] dm.$$

(3) The theorem of Jessen-Marcinkiewicz-Zygmund.

When one applies the previous remark iteratively one obtains the following result:

In R^n one considers the basis $\mathcal{B}^s$ such that $\mathcal{B}^s(x)$ for $x \in R^n$ is the collection of all open bounded intervals containing x such that s of the n side lengths are equal and the other n - s are arbitrary. Then one has for the corresponding maximal operator M^s and for each $\lambda > 0$, $f \in L_{loc}(R^n)$

$$|\{M^s f > \lambda\}| \leq c \int \frac{|f|}{\lambda} (1 + \log^+ \frac{|f|}{\lambda})^{n-s}$$

We leave the details as an exercise.

From this result and following exactly the process of the proof of the Lebesgue differentiation theorem given in Remark (3) of section 1, one obtains:

For each $f \in L(1 + \log^+ L)^{n-s}(R^n)$ (i.e. f is such that

$$\int |f|(1 + \log^+|f|)^{n-s} < \infty)$$

one has at almost every $x \in R^n$

$$\lim \frac{1}{|B|} \int_B |f(y) - f(x)| \, dy = 0$$

as $\delta(B) \to 0$, $B \in \mathcal{B}^s(x)$.

This theorem for s = 1 was obtained by Jessen, Marcinkiewicz and Zygmund [1935] and for s arbitrary by Zygmund [1967] with proofs different than the one presented here.

The first result of this remark, about the weak type inequality for the maximal operator M^s can also be obtained as a corollary of the Jessen-Marcinkiewicz-Zygmund theorem and of a result of Rubio [1972] that will be presented in a more general version in III.3.

4. THE ROTATION METHOD IN THE STUDY OF THE MAXIMAL OPERATOR.

For the study of the maximal operator associated to certain differentiation bases it is of interest the introduction of the rotation method developped by Calderón and Zygmund [1956] for the treatement of the classical singular integral operators.

The bases we consider in this section are of the following type: For the origin 0 of R^n, $\mathcal{B}(0)$ shall be a family of open bounded sets containing 0 symmetric and star-shaped with respect to 0, i.e. if $a \in B \in \mathcal{B}(0)$, then the whole segment $a\,0$ is in B. For any other point $x \in R^n$, $\mathcal{B}(x)$ is obtained translating the sets of $\mathcal{B}(0)$ by a parallel translation of vector x, i.e.,

$$\mathcal{B}(x) = \{x + B : B \in \mathcal{B}(0)\}$$

where $x + B = \{x + z : z \in B\}$.

In this case the maximal operator can be expressed in the following way. If $B \in \mathcal{B}(0)$ and $B_x = x + B \in \mathcal{B}(x)$, we can write for $f \in L_{loc}(R^n)$,

$$\frac{1}{|B_x|}\int_{y\in B_x}|f(y)|dy = \frac{1}{|B|}\int_{x-y\in B}|f(y)|dy = \frac{1}{|B|}\int|f(y)|\chi_B(x-y)dy =$$

$$= |f| * \frac{\chi_B}{|B|}(x) = |f| * \phi_B(x)$$

where we have called $\phi_B = \frac{\chi_B}{|B|}$. So, if M is the maximal operator of $\mathcal{B}$ we can set

$$Mf(x)) = \sup\{|f| * \phi_B(x) : B \in \mathcal{B}(0)\}$$

The following theorem is interesting and easy to prove by means of the rotation method.

<u>THEOREM</u> 4.1. <u>Let $\mathcal{B}$ the basis indicated above. Let for $\bar{y} \in \Sigma$, i.e.</u> $|\bar{y}| = 1$,

$$a(\bar{y}) = \sup\{\frac{1}{|B|}|\phi_B(\bar{y})|^n : B \in \mathcal{B}(0)\}$$

<u>where</u> ϕ_B <u>is the following function characterizing the boundary of</u> B,

$$\phi_B(\bar{y}) = \sup\{\lambda : \lambda\,\bar{y} \in B\}$$

<u>Then, for each</u> p <u>with</u> $1 < p < \infty$ <u>and each</u> $f \in L^p$ <u>one has with a constant</u> c, <u>independent of</u> $\mathcal{B}$, f, p,

$$||Mf||_p \leq \frac{c}{p-1} \left(\int_\Sigma a(\bar{y})d\bar{y}\right) ||f||_p.$$

<u>Proof</u>. Assume, for convenience of notation, $f \geq 0$. We can write

$$f * \phi_B(x) = \int f(x-y)\,\phi_B(y)dy = \int_\Sigma \int_0^\infty f(x-r\bar{y})\,\phi_B(r\bar{y})r^{n-1}dr\,d\bar{y}$$

We can write, for a fixed $\bar{y} \in \Sigma$,

$$M_{\bar{y}}\, f(x) = \sup \left\{\int_0^\infty f(x-r\bar{y})\,\phi_B(r\bar{y})r^{n-1}dr \,:\, B \in \mathcal{B}(0)\right\}$$

It is then clear that

$$Mf(x) = \sup \{f * \phi_B(x) \,:\, B \in \mathcal{B}(0)\} \leq \int_\Sigma M_{\bar{y}}\, f(x)\, d\bar{y}$$

If we can prove

$$||M_{\bar{y}}\, f||_p \leq \frac{c}{p-1}\, a(\bar{y})\, ||f||_p \qquad (*)$$

then, by Minkowski's integral inequality, we get

$$||Mf||_p \leq \int_\Sigma ||M_{\bar{y}}\, f(\cdot)||_p d\bar{y} \leq \frac{c}{p-1} \int_\Sigma a(\bar{y})d\bar{y}\,||f||_p$$

and then the theorem is proved.

In order to obtain (*) we try to use the strong type of the maximal operator for the basis of intervals $\mathcal{B}_1$ in R^1.

In

$$M_{\overline{y}}f(x) = \sup \{\int_0^\infty f(x-r\overline{y})\ \phi_B(r\overline{y})r^{n-1}\ dr : B \in \mathcal{B}(0)\}$$

we express $x = z + s\overline{y}$ with $z \in H$, H being the hyperplane passing through $\overline{y}$ orthogonal to $\overline{y}$, and $s \in R$. Since $\overline{y}$ is fixed we write $\phi_B(r\overline{y}) = \eta_B(r)$. Hence

$$M_{\overline{y}}f(z+s\overline{y}) = \sup \{\int_0^\infty f(z+(s-r)\overline{y})\eta_B(r)r^{n-1}dr : B \in \mathcal{B}(0)\}$$

Let us now fix also $z \in H$ and call

$$f(z + t\overline{y}) = g(t)$$

Observe that $\eta_B(r) = \frac{1}{|B|}$ if $0 < r < \phi_B(\overline{y})$ and $\eta_B(r) = 0$ if $r \geq \phi_B(\overline{y})$. Therefore,

$$M_{\overline{y}}f(z+s\overline{y}) = \sup \{\frac{1}{|B|}\int_0^{\phi_B(\overline{y})} g(s-r)\ r^{n-1}\ dr : B \in \mathcal{B}(0)\} =$$

$$= \sup \{\frac{(\phi_B(\overline{y}))^n}{|B|}\ \frac{1}{\phi_B(\overline{y})}\int_0^{\phi_B(\overline{y})} g(s-r)\ [\frac{r}{\phi_B(\overline{y})}]^{n-1}\ dr : B \in \mathcal{B}(0)\}$$

Since the expression in brackets is less than or equal to 1, we have

$$M_{\overline{y}}\ f(z + s\overline{y}) \leq a(\overline{y})\ M^*g(s)$$

where we have called

$$M^*g(x) = \sup \{\frac{1}{h}\int_0^h g(s-r)\ dr : h > 0\}$$

Because of the strong type (p,p) for the maximal operator of the basis of intervals in R^1 (cf. Remark (2) of Section 1), we have

$$\int |M^*\ g(r)|^p\ dr \leq \frac{c^p}{(p-1)^p}\int |g(r)|^p\ dr$$

So we can write

$$||M_{\overline{y}} f(\cdot)||_p^p = \int_{z\in H} \int_{s\in R} |M_{\overline{y}} f(z+s\overline{y})|^p \, ds \, dz \leq$$

$$\leq \frac{c^p}{(p-1)^p} (a(\overline{y}))^p \int_{z\in H} \int_{s\in R} |f(z+s\overline{y})|^p \, ds \, dz =$$

$$= \frac{c^p}{(p-1)^p} (a(\overline{y}))^p \, ||f||_p^p.$$

This concludes the proof of the theorem.

REMARKS.

(1) An application of the theorem to homothetic sets.

If all sets of $\mathcal{B}(0)$ are homothetic to a fixed set B then clearly $\frac{(\phi_B(\overline{y}))^n}{|B|}$ is a constant independent of B and $\overline{y}$ and so $a(\overline{y})$ is constant. So we get

$$||Mf||_p \leq \frac{c}{p-1} ||f||_p$$

By the process of the proof of the Lebesgue differentiation theorem given in Remark (3) of Section 1, we have here for each $f \in L^p$, with $1 < p < \infty$, and almost all $x \in R^n$

$$\lim \frac{1}{|B|} \int_B |f(y) - f(x)| \, dy = 0$$

as $\delta(B) \to 0$, $B \in \mathcal{B}(x)$.

The rotation method, as we can see gives us the strong type (p,p), $1 < p \leq \infty$ of the maximal operator of a basis of homothetic sets as the one considered here, but not the weak type (1,1). This one, which is valid also in this case, is obtained by means of covering theorems in R^n that the rotation method avoids.

(2) The theorem 3.1. is valid for unbounded sets.

Observe that the fact that the sets of $\mathcal{B}(0)$ are bounded is rather irrelevant for the proof of Theorem 3.1. Assume that $\mathcal{B}(0)$ is formed by open sets symmetric and starshaped with respect to 0 and of finite measure. As before, we define

$$Mf(x) = \sup \{ |f| * \phi_B(x) : B \in \mathcal{B}(0)\}$$

Then, if the function $a(\overline{y})$ is in $L^1(\Sigma)$ (for instance, if all sets in $\mathcal{B}(0)$ are homothetic to a fixed one) we obtain

$$||Mf||_p \leq \frac{c}{p-1} ||f||_p$$

(3) The role of the constant $\frac{c}{p-1}$ in Theorem 3.1.

The constant $\frac{c}{p-1}$ in the strong type (p,p), $1 < p < \infty$ of the maximal operator is important. As we shall see in Chapter VI, by means of a method of extrapolation due to Yano, one can obtain a stronger inequality using this value of the constant.

5. A CONVERSE INEQUALITY FOR THE MAXIMAL OPERATOR.

For the maximal operator associated to the basis $\mathcal{B}_1$ of open cubic intervals of R^n there is an inequality that presents some interest in itself and also for the characterizations of some important spaces in differentiation and in Analysis in general. Such is the space $L(1 + \log^+ L)$ of Zygmund, whose characterization is given in Section 6. An inequality of the type described here appears in Stein [1969] and Herz [1968]. The proof of this fact we present here, published here for the first time, is based on Whitney's covering theorem and permits an improvement of their results.

THEOREM 5.1. *Let $\mathcal{B}_1$ be the basis such that $\mathcal{B}_1(x)$ is the collection of all open bounded cubic intervals of R^n containing x and let M_1 be its maximal operator*

Then, for each $f \in L(R^n)$ and for each $\lambda > 0$ we have

$$\frac{c_n^*}{\lambda}\int_{M_1f>\lambda}|f| \leq |\{M_1f > \lambda\}| \leq \frac{c_n}{\lambda}\int_{|f|>\frac{\lambda}{2}}|f|$$

where c_n, c_n^* are constants depending only on n.

Proof. The second inequality is merely the refinement of the weak type (1,1) for M_1 given in Remark (1) of Section 1.

The first inequality is obtained in the following way. Let $A = \{M_1 f > \lambda\}$. The set A is open. If $A = \phi$ the inequality is trivial. The fact that $f \in L^1(R^n)$ easily implies that $A \neq R^n$. So we can apply the theorem of Whitney of I.2. obtaining a sequence $\{Q_k\}$ of half-open disjoint cubic intervals such that $A = \bigcup Q_k$ and

$$1 \leq \frac{d(Q_k, \partial A)}{\delta(Q_k)} \leq 3.$$

If we set $Q_k^* = a_n Q_k$, i.e. Q_k^* is the expansion by a_n of Q_k with center at the center of Q_k, then it is clear that choosing conveniently the constant a_n, that depends only on the dimension, we have

$$Q_k^* \cap A' \neq \phi \text{ and so } \frac{1}{|Q_k^*|}\int_{Q_k^*}|f| \leq \lambda.$$

Hence, if we call $(\frac{1}{a_n})^n = c_n^*$

$$|A| = \sum_k |Q_k| = c_n^* \sum |Q_k^*| \geq \frac{c_n^*}{\lambda}\sum_k \int_{Q_k^*}|f| \geq \frac{c_n^*}{\lambda}\sum_k \int_{Q_k}|f| = \frac{c_n^*}{\lambda}\int_A |f|$$

This is the inequality we wanted.

REMARKS.

(1) The converse inequality for more general bases.

The inequality we have proved is valid also for any differentiation basis that satisfies even the Whitney's theorem in the weaker form of Theorem I.2.2., i.e. with disjointness replaced by uniformly bounded overlap. We leave the details as an exercise.

(2) A finer theorem in R^1.

By using the idea of the wellknown rising sun lemma of F. Riesz [1932] (cf. also Boas [1960, p.134]) we can obtain in R^1 a result finer than Theorem 5.1.

For $f \in L^1(R^1)$ we define the following maximal operator

$$Mf(x) = \sup \{ \frac{1}{u-x} \int_x^u |f| : u > x \}$$

Then, for each $\lambda > 0$, the set $A = \{Mf > \lambda\}$ is either empty or else $A = \bigcup_k I_k$ with $I_k = (a_k, b_k)$, $I_k \cap I_j = \phi$ if $k \neq j$, $-\infty < a_k < b_k < +\infty$ and

$$\lambda = \frac{1}{b_k - a_k} \int_{a_k}^{b_k} |f|.$$

Hence, in any case,

$$|\{Mf > \lambda\}| = \frac{1}{\lambda} \int_{Mf>\lambda} |f|.$$

For the proof lets us first observe that the set $A = \{Mf > \lambda\}$ is open by the continuity of the integral with respect to the limits of integration. Hence, if $A \neq \phi$, then $A = \bigcup I_k$ where $\{I_k\}$ is a disjoint sequence of open intervals. Let us write $I_k = (a_k, b_k)$. That $-\infty < a_k < b_k < +\infty$ for each k is an easy consequence of the fact that $f \in L^1$. We shall prove that

$$\int_{a_k}^{b_k} |f| = \lambda(b_k - a_k)$$

for each k. If

$$\int_{a_k}^{b_k} |f| > \lambda(b_k - a_k),$$

then clearly $a_k \in A$ and this is impossible. Assume that

$$\int_{a_k}^{b_k} |f| > \lambda(b_k - a_k).$$

Take $\varepsilon > 0$ such that $a_k + \varepsilon \in (a_k, b_k)$ and let

$$p(\varepsilon) = \sup \{u \in R^1 : \frac{1}{u-(a_k+\varepsilon)} \int_{a_k+\varepsilon}^{u} |f| > \lambda\}$$

Then we have

$$\int_{a_k+\varepsilon}^{p(\varepsilon)} |f| \geq \lambda(p(\varepsilon) - (a_k + \varepsilon)) \qquad (*)$$

If ε is sufficiently small, then

$$\int_{a_k+\varepsilon}^{b_k} |f| < \lambda(b_k - (a_k+\varepsilon)) \qquad (**)$$

and so $p(\varepsilon) \neq b_k$.

If $p(\varepsilon) \in (a_k, b_k)$ then there would be $q > p(\varepsilon)$ such that

$$\int_{p(\varepsilon)}^{q} |f| > \lambda(q - p(\varepsilon)) \qquad (***)$$

and so, adding up (*) and (***),

$$\int_{a_k}^{q} |f| > \lambda(q - (a_k + \varepsilon))$$

and this contradicts the definition of $p(\varepsilon)$. Hence $p(\varepsilon) > b_k$ and so, subtracting (**) from (*), we get

$$\int_{b_k}^{p(\varepsilon)} |f| > \lambda(p(\varepsilon) - b_k).$$

This implies $b_k \in A$, which is contradictory. Thus we have proved

$$\int_{a_k}^{b_k} |f| = \lambda(b_k - a_k).$$

6. THE SPACE $L(1 + \log^+ L)$. INTEGRABILITY PROPERTIES OF THE MAXIMAL OPERATOR.

The space of all measurable functions $f : R^n \to R$ such that

$$\int |f| \; (1 + \log^+ |f|) < \infty$$

and other similar spaces play an important role in many problems in Analysis. This kind of spaces was introduced by Zygmund [1929] for certain problems in Fourier analysis. We shall see in this section how the functions f in $L(1 + \log^+ L)$ can be characterized by means of the maximal operator. This characterization appears in Guzmán and Welland [1971]. A previous result in this direction is due to Stein [1969].

THEOREM 6.1. *Let* $f \in L^1(R^n)$. *Then the two following conditions are equivalent:*

i) $\displaystyle\int_{M_1 f > 1} M_1 f < \infty$

ii) $f \in L(1 + \log^+ L)$

where M_1 *is the maximal operator associated to the basis* $\mathcal{B}_1$ *of open cubic intervals.*

Proof. We can assume, without loss of generality, that $f \geq 0$. We shall make use of the weak type (1,1) of M_1 and of the converse inequality already proved in the previous section. In the course of the proof c will denote a positive constant, not necessarily the same at each ocurrence, depending only on n. We make use of the following expression of the integral of a measurable function $g \geq 0$, defined on R^n

$$\int g(x)\,dx = -\int_0^\infty \lambda dF(\lambda)$$

where, for $\lambda > 0$, $F(\lambda) = |\{x \in R^n : g(x) > \lambda\}|$. The function F so defined is called the distribution function of g.

Assume that f satisfies i). Then

$$\int_{M_1 f>1} M_1 f = -\int_1^\infty \lambda d\omega(\lambda) = \left[-\lambda\omega(\lambda)\right]_1^\infty + \int_1^\infty \omega(\lambda)\,d\lambda$$

where

$$\omega(\lambda) = |\{x : M_1 f(x) > \lambda\}|.$$

Since

$$|\{x : M_1 f(x) > 1\}| \leq c||f||_1$$

and

$$\lambda|\{x \in R^n : M_1 f(x) > \lambda\}| \leq \int_{f>\frac{\lambda}{2}} f \to 0 \text{ for } \lambda \to \infty,$$

we get, using the converse inequality of Section 5, (cf. Theorem 5.1),

$$\infty > \int_1^\infty \omega(\lambda)d\lambda \geq c\int_1^\infty \frac{1}{\lambda}\int_{M_1 f(x)>\lambda} f(x)dxd\lambda \geq$$

$$\geq c\int_1^\infty \frac{1}{\lambda}\int_{f(x)>\lambda} f(x)dxd\lambda = c\int_{f(x)>1} f(x)\int_1^{f(x)} \frac{1}{\lambda}\,d\lambda dx =$$

$$= c\int f(x)\log^+ f(x)\,dx$$

This, since $f \in L$, implies $f \in L(1 + \log^+ L)$.

Let us now assume that ii) holds. We then have as before

$$\int_{M_1 f>1} M_1 f = \omega(1) + \int_1^\infty \omega(\lambda)\, d\lambda$$

Now $\omega(1) = |\{M_1 f > 1\}| \leq c||f||_1 < \infty$ and

$$\int_1^\infty \omega(\lambda)d\lambda \leq c\int_1^\infty \frac{1}{\lambda}\int_{f(x)>\frac{\lambda}{2}} f(x)dx\, d\lambda =$$

$$= c\int_{2f(x)>1}\int_1^{2f(x)} \frac{1}{\lambda}\, d\lambda dx = c\int_{2f>1} f \log^+ 2f < \infty.$$

So ii) $\Rightarrow$ i). This concludes the proof of the theorem.

REMARKS.

(1) The theorem of Stein.

The first characterization of the type we have presented here appears in Stein [1969]. He proved a result which can be formulated in the following way: Let $f \in L^1(R^n)$ and with compact support contained in a closed cubic interval Q. Then $f \in L \log^+ L$ if and only if $\int_Q M_1^* f < \infty$, where M_1^* is the maximal operator associated to $\mathcal{B}_1^*$, the basis such that $\mathcal{B}_1^*(x)$ is the collection of open cubic intervals centered at x.

This result can be obtained as an easy consequence of Theorem 6.1. which is obviously valid for this basis $\mathcal{B}_1^*$. If $f \in L \log^+ L$ and has compact support, then $f \in L(1 + \log^+ L)$ and so, according to the theorem

$$\int_{M_1^* f>1} M_1^* f < \infty.$$

But

$$\int_Q M_1^* f = \int_{Q\cap\{M_1^* f>1\}} M_1^* f + \int_{Q\cap\{M_1^* f\leq 1\}} M_1^* f \leq \int_{M_1^* f>1} M_1^* f + |Q| < \infty$$

Assume now that $\int_Q M_1^*f < \infty$. Let us prove first that M_1^*f is locally integrable. For this we take the cube Q^* indicated in Figure 1:

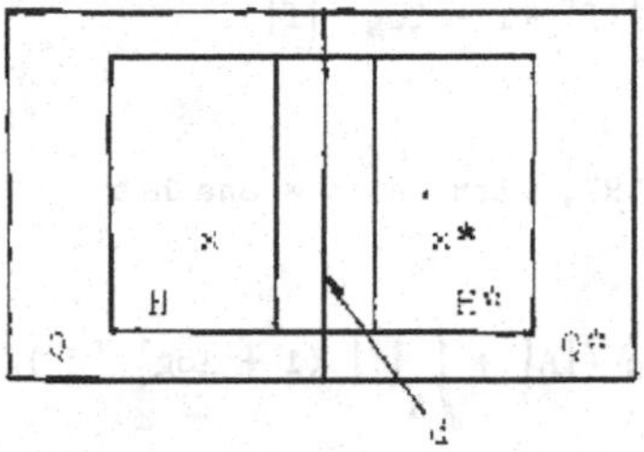

Fig. 1

To each $x^* \in Q^*$ we make correspond the point x, symmetric with respect to d. To each cube $H^* \in \mathcal{B}_1^*(x^*)$ we make correspond the symmetric $H \in \mathcal{B}_1^*(x)$ with respect to d. Since the support of f is in Q,

$$\frac{1}{|H^*|}\int_{H^*}|f| = \frac{1}{|H^*|}\int_{H^*\cap Q}|f| \leq \frac{1}{|H|}\int_H |f|$$

and so $M_1^*f(x^*) \leq M_1^*f(x)$. From this we easily conclude

$$\int_{Q^*} M_1^*f \leq \int_Q M_1^*f < \infty.$$

Following the same idea we get $M_1^*f \in L_{loc}(R^n)$ as we wanted.

If we now prove that $\{M_1^*f > 1\}$ is bounded we have $\int_{M_1^*f>1} M_1^*f < \infty$ and so, by the theorem 6.1., $f \in L(1 + \log^+ L)$, and a fortiori $f \in L \log^+ L$. If $x \in R^n$ is far away from Q and $H \in \mathcal{B}_1^*(x)$ is such that $H \cap Q \neq \phi$, the ratio $\frac{1}{|H|}\int_H |f|$ is less than

$$\frac{c||f||}{|d(x,Q)|^n}$$

c being a constant that depends on the dimension. So $\{M_1^*f > 1\}$ is bounded and this concludes the proof.

(2) Another inequality.

The following inequality can be easily proved by the same method used in the theorem 6.1.: Let $f \in L^1(R)$. Then

$$\int M_1^* f \geq c \int |f| \, (1 + \log^+ |f|).$$

Also, for each measurable set $A \subset R^n$, with $|A| < \infty$ one has

$$\int_A M_1^* f \leq c \, (|A| + \int_A |f| \, (1 + \log^+ |f|).$$

The reader is invited to prove this fact or else he is referred to Guzmán and Welland [1971].

(3) <u>The space</u> $L(1 + \log^+ L)^2(R^2)$.

If $f \in L(1 + \log^+ L)^2 \, (R^2)$ and M_2 is the maximal operator associated to $\mathcal{B}_2$, the basis of intervals in R^2, then one has

$$\int_{M_2 f > 1} M_2 f < \infty.$$

In fact, we know from Section 3 that, for each $\lambda > 0$,

$$|\{M_2 f > \lambda\}| \leq c \int \frac{|f|}{\lambda} \, (1 + \log^+ \frac{|f|}{\lambda}).$$

From this inequality one can proceed as in Theorem 6.1. to obtain

$$\int_{M_2 f > 1} M_2 f < \infty.$$

It would be interesting to know whether the space $L(1 + \log^+ L)^2(R^2)$ can be characterized by means of M_2 as we have done with $L(1 + \log^+ L)$.

CHAPTER III

THE MAXIMAL OPERATOR AND THE DIFFERENTIATION PROPERTIES OF A BASIS

The notion of derivation of the integral of a function with respect to a differentiation basis $\mathcal{B}$ is a natural extension of the local derivation in the sense of the theorem of Lebesgue.

Let $f \in L_{loc}(R^n)$. For $x \in R^n$ we define the <u>upper derivative</u> of f with respect to $\mathcal{B}$ at x in the following way

$$\overline{D}\left(\int f,x\right) = \sup \left\{\limsup_{k\to\infty} \frac{1}{|B_k|}\int_{B_k} f : \{B_k\}\subset\mathcal{B}(x),\ B_k \to x\right\}$$

and similarly the <u>lower derivative</u>

$$\underline{D}\left(\int f,x\right) = \inf \left\{\liminf_{k\to\infty} \frac{1}{|B_k|}\int_{B_k} f : \{B_k\} \quad (x),\ B_k \to x\right\}$$

We say that $\mathcal{B}$ <u>differentiates</u> f when we have

$$\overline{D}\left(\int f,x\right) = \underline{D}\left(\int f,x\right) = f(x)$$

at almost every $x \in R^n$. In this case we denote

$$\overline{D}\left(\int f,x\right) = D\left(\int f,x\right) = \underline{D}\left(\int f,x\right).$$

When $\mathcal{B}$ differentiates $\int f$ for each f in a class X of functions we shall also say, somewhat improperly, that $\mathcal{B}$ <u>differentiates</u> X.

The main problem we shall treat in this chapter consists in studying properties of the maximal operator bearing on the differentiability properties of the corresponding basis $\mathcal{B}$. More specifically if $\mathcal{B}$ differentiates for example L^∞, can anything be said about the behavior of the maximal operator associated to $\mathcal{B}$?

In all considerations of this chapter, mainly for the sake of brevity, $\mathcal{B}$ will always be a Busemann-Feller basis (B-F basis). We refer to II.2 for its definition. We have seen that for such a basis, if $f \in L_{loc}(R^n)$, Mf is always measurable. It is also easy to prove that for a B-F basis and for a function $f \in L_{loc}(R^n)$ the functions $\overline{D}(\int f, \cdot)$, $\underline{D}(\int f, \cdot)$ are always measurable. In fact, for any $a \in R$ we have

$$A = \{x \in R^n : \overline{D}(f,x) > a\} = \bigcup_{r=1}^{\infty} \bigcap_{s=1}^{\infty} A_{rs}$$

Where for each, r, s,

$$A_{rs} = \{B \in \mathcal{B}(x) : \delta(B) < \frac{1}{s}, \frac{1}{|B|} \int_B f \geq a + \frac{1}{r}\}.$$

Since all sets $B \in \mathcal{B}(x)$ are open, so is A_{rs}. Hence $\overline{D}(\int f, \cdot)$ is measurable. Since

$$\underline{D}(\int f,x) = -\overline{D}(\int (-f),x),$$

also $\underline{D}(\int f, \cdot)$ is measurable.

1. DENSITY BASES. THEOREMS OF BUSEMANN-FELLER.

We shall say that a B-F basis $\mathcal{B}$ satisfies the density property or is a density basis is for each measurable set A we have $D(\int \chi_A, x) = \chi_A(x)$ at almost every $x \in R^n$.

This can also be expressed by saying that for each measurable set A, for almost every $x \in R^n$ we have, if $\{R_k\}$ is any arbitrary sequence of $\mathcal{B}(x)$ contracting to x,

$$\lim_{k \to \infty} \frac{|A \cap R_k|}{|R_k|} = \chi_A(x)$$

This accounts better for the reference to the density.

We now present two theorems, essentially due to Bussemann and Feller [1934] characterizing the density bases in terms of the maximal operator. The explicit reference to the maximal operator is due to Guzmán and Welland [1971] and lends itself to some useful and interesting considerations, as we shall later see.

THEOREM 1.1. *Let $\mathcal{B}$ be a B-F basis. The two following properties are equivalent:*

(a) *$\mathcal{B}$ is a density basis.*

(b) *For each λ, $0 < \lambda < 1$, for each nonincreasing sequence $\{A_k\}$ of bounded measurable sets such that $|A_k| \downarrow 0$ and for each nonincreasing sequence $\{r_k\}$ of real numbers such that $r_k \downarrow 0$ we have*

$$|\{M_k\chi_k > \lambda\}| \to 0$$

where, for each k, h, $\chi_h = \chi_{A_h}$, and

$$M_k\chi_h(x) = \sup \{\frac{1}{|B|}\int_B \chi_h : \delta(B) < r_k,\ B \in \mathcal{B}(x)\}$$

Proof. That (a) implies (b) is easy. Let $0 < \lambda < 1$ and $\{A_k\}$ as in (b). Fix an A_h. For almost each $x \notin A_h$ we have, if (a) is true, $D(\int \chi_h, x) = 0$, and so, if k is sufficiently big, $M_k\chi_h(x) \leq \lambda$. Hence

$$\overline{\lim_{k\to\infty}}\ |\{M_k\chi_h > \lambda\}| \leq |A_h|.$$

For each $k \geq h$ one has

$$\{M_k\chi_k > \lambda\} \subset \{M_k\chi_h > \lambda\}$$

by the definition of $M_k\chi_h$. Therefore

$$\overline{\lim_{k\to\infty}}\ |\{M_k\chi_k > \lambda\}| \leq |A_h|.$$

Since $|A_h| \to 0$ we get (b).

We now prove that not-(a) implies not-(b). If $\mathcal{B}$ is not a density basis, there is a measurable set A, with $|A| > 0$, such that

$$\left|\{x \notin A : \overline{D}(\int \chi_A, x) > 0\}\right| > 0.$$

In fact, assume that for each measurable set $|P| > 0$, we have, at almost each $x \notin P$, $\overline{D}(\int \chi_P, x) = 0$, i.e. $\overline{D}(\int \chi_P, x) = 0 = \underline{D}(\int \chi_P, x)$. If we apply this to the complement P' of P, if $|P'| > 0$, we obtain that at almost each $x \notin P'$, i.e. at almost each $x \in P$, we have

$$\overline{D}(\int \chi_{P'}, x) = 0 = \underline{D}(\int \chi_{P'}, x)$$

Observe now that

$$\overline{D}(\int \chi_{P'}, x) = 1 - \underline{D}(\int \chi_P, x)$$

$$\underline{D}(\int \chi_{P'}, x) = 1 - \overline{D}(\int \chi_P, x)$$

and so we have, at almost each $x \in P$

$$\overline{D}(\int \chi_P, x) = 1 = \underline{D}(\int \chi_P, x)$$

and therefore $\mathcal{B}$ would be a density basis.

Let us then take A measurable, with $|A| > 0$, such that

$$\left|\{x \notin A : \overline{D}(\int \chi_A, x) > 0\}\right| > 0.$$

There exists then a measurable set C, with $C \subset A'$, $|C| > 0$, such that at each $x \in C$ we have $\overline{D}(\int \chi_A, x) > \lambda$. Let $\{G_k\}$ a sequence of nonincreasing open sets such that $G_k \supset C$, $|G_k - C| \to 0$ and let $A_k = G_k \cap A$. Clearly $\{A_k\}$ is nonincreasing and $|A_k| \to 0$

since $A_k \subset G_k - C$. Take any nonincreasing sequence of real numbers $\{r_k\}$ such that $r_k \to 0$. We shall prove that $\{M_k\chi_{A_k} > \lambda\} \supset C$ for each k. In fact, let $x \in C$ and k be fixed. Since $\overline{D}(\int \chi_A, x) > \lambda$ there is a sequence $\{R_h\} \subset \mathcal{B}(x)$ with $R_h \to x$ such that $R_h \subset$ $\subset G_k$

$$\frac{|R_h \cap A|}{|R_h|} > \lambda .$$

Hence

$$\frac{|R_h \cap A_k|}{|R_h|} > \lambda$$

and $M_k\chi_{A_k}(x) > \lambda$. This proves $C \subset \{M_k\chi_k(x) > \lambda\}$ for each k and, since $|C| > 0$, this shows that not-(b) holds. This concludes the proof of the theorem.

When $\mathcal{B}$ is a B-F basis that is invariant by homothecies, i.e. when $\mathcal{B}$ is such that if $R \in \mathcal{B}$ then any set homothetic to R with any ratio and any center of homothecy is also in $\mathcal{B}$ (and so in particular any translated set of R), then the preceding criterion receives a simpler form, as the following theorem proves.

<u>THEOREM</u> 1.2. <u>Let $\mathcal{B}$ be a B-F basis that is invariant by homothecies. Then the two following properties are equivalent:</u>

(a) <u>$\mathcal{B}$ is a density basis.</u>

(b) <u>For each λ, $0 < \lambda < 1$, there exists a positive constant</u> $c(\lambda) < \infty$ <u>such that for each bounded measurable set</u> A <u>one has</u>

$$|\{M\chi_A > \lambda\}| \leq c(\lambda)\, |A| .$$

<u>Proof</u>. That (b) implies (a) is a simple consequence of the theorem 1.1., since (b) implies condition (b) of that theorem.

In order to prove that (a) implies (b) we shall use the following lemma.

LEMMA 1.3. Let G be any bounded open set in R^n and let K be any compact set with positive measure. Let $r > 0$. Then there is a disjoint sequence $\{K_k\}$ of sets homothetic to K contained in G such that $|G - \bigcup K_k| = 0$ and $\delta(K_k) < r$.

Proof of the lemma. The lemma is an easy consequence of the fact the basis $\mathcal{B}_K$ of all sets homothetic to K satisfies the theorem of Vitali. However a simple proof of it can be given in the following way.

Let A be a half-open cubic interval such that $K \subset \overset{\circ}{A}$ and let $\alpha|A| = |K|$ with $0 < \alpha < 1$. We partition G into a sequence of disjoint half-open cubic intervals $\{A_h\}$ of diameter less than r. For each A_h let p_h be the homothecy that carries A onto A_h and let $K_h^* = p_h K$. We can keep a sequence $\{K_h^*\}_{h=1}^{N_1}$ of these sets such that, if

$$G_1 = G - \bigcup_1^{N_1} K_h^*,$$

then G_1 is open and

$$|G_1| = |\bigcup_1^{\infty} A_h - \bigcup_1^{N_1} K_h^*| = |\bigcup_{h=1}^{N_1} (A_h - K_h^*)| + |\bigcup_{h>N_1} A_h| = (1-\alpha)|\bigcup_{h=1}^{N_1} A_h| +$$

$$+ |\bigcup_{h>N_1} A_h| < (1 - \frac{\alpha}{2})\, |G|,$$

by taking N_1 sufficiently big. We now set $K_h = K_h^*$, $h = 1, 2, \ldots, N_1$ and proceed with G_1 as we have done with G, obtaining now $\{K_h\}_{h=N_1+1}^{N_2}$ such that

$$|G_2| = |G_1 - \bigcup_{h=N_1+1}^{N_2} K_h| \leq (1-\frac{\alpha}{2})|G_1| \leq (1-\frac{\alpha}{2})^2|G|.$$

And so on. So we obtain the sequence $\{K_h\}$ satisfying the lemma.

We now continue with the proof of the theorem. Assume that (b) does not hold. The there exists a positive number $\lambda > 0$ such that for each integer $k > 0$ there is a bounded measurable set A_k such that, if $\chi_k = \chi_{A_k}$,

$$|\{M\chi_k > \lambda\}| > 2^{k+1} |A_k|$$

There is also a positive number r_k such that

$$|\{M_k \chi_k > \lambda\}| > 2^{k+1} |A_k|$$

where M_k means M_{r_k}. Let C_k be a compact subset of $\{M_k \chi_j > \lambda\}$ such that $|C_k| > 2^{k+1}|A_k|$. By the previous lemma we can cover the open unit cube Q almost completely by means of a disjoint sequence $\{C_k^j\}_{i=1,2,...}$ of sets homothetic to C_k such that if α_{kj} is the ratio of the homothecy p_{kj} carrying C_k onto C_k^j we have $\alpha_{kj} r_k < 2^{-k}$ for each j and k. Let $p_{kj} A_k = A_k^j$ and let A be the union of all sets A_k^j, k = 1, 2, ..., j = 1, 2, ... We then have

$$|A| \leq \sum_{j,k} |A_k^j| < \sum_{k,j} 2^{-(k+1)} |C_k^j| = \sum_{k=1}^{\infty} 2^{-(k+1)} \sum_j |C_k^j| = \frac{1}{2}$$

We shall now prove that at almost each $x \in Q$ we have $\overline{D}(\int \chi_A, x) \geq \lambda > 0$. Since $|A| < \frac{1}{2}$ this will prove that the density property is not true for A.

Fix k and let $x \in C_k$. There is then $R \in \mathcal{B}(x)$, with $\delta(R) < r_k$ such that

$$\frac{|R \cap A_k|}{|R|} > \lambda$$

For each j, the image R^* of R by the homothecy p_{k_j} is such that $\delta(R^*) < 2^{-k}$ and

$$\frac{|R^* \cap A|}{|R^*|} > \lambda .$$

Since for each fixed k almost every point x of Q is in some C_k^j, it results that for almost each x of Q there is a sequence R_k of elements of $\mathcal{B}(x)$ contracting to x such that

$$\frac{|R_k \cap A|}{|R_k|} > \lambda .$$

Thus $\overline{D}(\chi_A, x) \geq \lambda$ almost everywhere in Q.

The following theorem affirms that the density property is in fact equivalent to another apparently stronger property, that of differentiating L^∞. This result belongs to Busemann and Feller [1934]. The proof presented here is a little different.

THEOREM 1.4. Let $\mathcal{B}$ be a density basis. Then $\mathcal{B}$ differentiates L^∞.

Proof. Since the differentiation of $\int f$ at x is a local property, i.e. depends only on the behavior of f in a neighborhood of x we can assume that f has compact support A. We also can assume without loosing generality that for every x, $0 \leq f(x) \leq H < \infty$. By Lusin's theorem, given $\varepsilon > 0$, there exists a compact set K in A such that $|A - K| \leq \varepsilon$ and f is continuous on K. Let $f_K = f\chi_K$, $f_{A-K} = f\chi_{A-K}$.

We first prove $D(\int f_K, x) = f_K(x)$ at almost every $x \in R^n$. In fact, assume $R_k \in \mathcal{B}(x)$, $R_k \to x$ as $k \to \infty$. We can write

$$\left|\frac{1}{|R_k|}\int_{R_k} f_K(y)dy - f_K(x)\right| \leq \frac{1}{|R_k|}\int_{R_k} |f_K(y) - f_K(x)|\, dy =$$

$$= \frac{1}{|R_k|}\int_{R_k\cap K} |f_K(y) - f_K(x)|dy + \frac{1}{|R_k|}\int_{R_k-K} |f_K(y)-f_K(x)|dy \leq$$

$$\leq \frac{1}{|R_k|}\int_{R_k\cap K} |f_K(y) - f_K(x)| + H\,\frac{|R_k-K|}{|R_k|}$$

If $x \in K$, then $f_K(y) \to f_K(x)$ as $y \to x$, $y \in K$ and so the above expression tends to zero as $k \to \infty$. If $x \notin K$, then $f_K(x) = 0$ and the first member is majorized by $\frac{|R_k\cap K|}{|R_k|}$. This, by the density property tends to zero for almost all such points x. Hence $D(\int f_K, x) = f_K(x)$ almost everywhere in R^n.

With this, for an arbitrary $\alpha > 0$, we can set

$$|\{x : |\overline{D}(\int f, x) - f(x)| > \alpha\}| = |\{x : |\overline{D}(\int f_{A-K}, x) - f_{A-K}(x)| > \alpha\}| \leq$$

$$\leq |A - K| + |\{x \in K : \overline{D}(\int f_{A-K}, x) > \alpha\}|$$

But $|A - K| \leq \varepsilon$ and for almost each $x \in K$ and for each $R_k \in \mathcal{B}(x)$, $R_k \to x$, we have,

$$\frac{1}{|R_k|}\int_{R_k} f_{A-K}(y)\, dy \leq H\, \frac{|R_k \cap (A-K)|}{|R_k|} \to 0$$

by the density property. So $\overline{D}(\int f,x) = f(x)$ for almost each $x \in R^n$. Similarly $\underline{D}(\int f,x) = f(x)$ almost everywhere and so the theorem is proved.

REMARKS.

(1) A basis not satisfying the density property.

By means of Theorem 1.2. it is very easy to produce a basis not having the density property. Let $\mathcal{B}$ be the basis of all open sets in R^2 with the shape of a solid Γ, i.e. such as the one of Fig. 2 with a, b, c, d > 0 arbitrary, and all other obtained from these ones

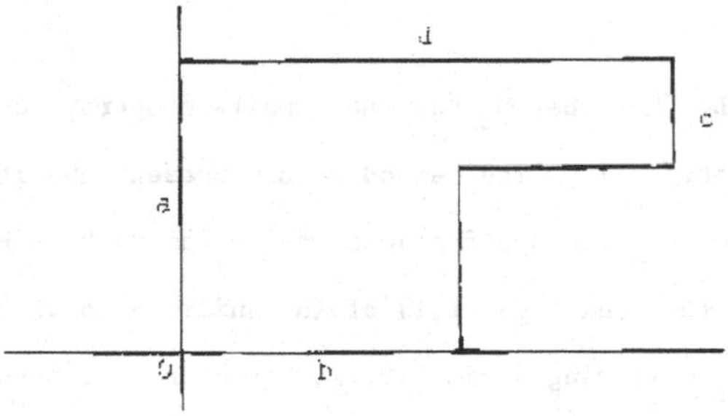

Fig. 2

by translations. Then if Q is the unit cube we have easily

$$|\{M\chi_Q > \tfrac{1}{2}\}| = \infty$$

since, as the Fig. 3 suggests, all points (x,y) with $0 < y < 1$ are in some $R \in \mathcal{B}$ with $\frac{|R \cap Q|}{|R|} > \frac{1}{2}$.

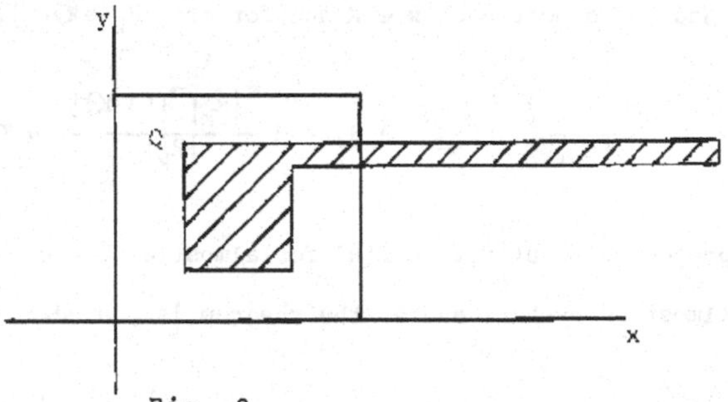

Fig. 3

(2) An easier proof that $\mathcal{B}_2$ is a density basis.

The inequality we have obtained for M_2 in II.3 together with Theorem 1.2. gives an immediate proof that $\mathcal{B}_2$ is a density basis. In fact, for each bounded measurable set A in R^n we have for $0 < \lambda < 1$

$$|\{M_2\chi_A > \lambda\}| \leq \frac{c}{\lambda}\left(1 + \log \frac{1}{\lambda}\right)^{n-1} |A|.$$

An easier proof of the fact that $\mathcal{B}_2$ has the density property can be obtained by means of a less difficult inequality. The method we now present has its origin in F. Riesz [1934], as one of the multiple applications of his famous "rising sun lemma". Here we shall make use of the weak type (1,1) of the unidimensional maximal operator with respect to intervals, following Rubio [1972], For clarity's sake we present here the proof for n = 1. It is to be observed the similarity of the argument with the one appearing in II.3.1. However, everything here is much simpler because one has to deal only with characteristic functions of sets.

For a set $P \subset R^2$ and for each x, y e R we define the set

$$P_2(x) = \{t \in R : (x,t) \in P\}$$

$$P_1(y) = \{t \in R : (t,y) \in P\}$$

i.e. $P_2(x)$ is the section of P by the straight line through (x,0) parallel to the axis Ox_2. Similarly for $P_1(y)$.

Let A be a bounded open set. We consider for a fixed λ, $0 < \lambda < 1$, the sets

$$B = \{(x,y) \in R^2 : M^*\chi_{A_2(x)}(y) > \lambda\}$$

$$C = \{(x,y) \in R^2 : M^*\chi_{B_1(y)}(x) > \lambda\}$$

where M^* means the unidimensional maximal operator with respect to open intervals i.e. if $f \in L(R)$, then

$$M^*f(t) = \sup \{\frac{1}{|I|} \int_I |f(s)|ds : I \text{ open interval of } R, t \in I\}.$$

It is easy to show that, since A is open, also B and C are open.

Because of the weak type (1,1) for M^*, we have for each fixed $x \in R$

$$|B_2(x)|_1 \leq \frac{c}{\lambda} |A_2(x)|_1$$

Where $| \ |_1$ means one-dimensional Lebesgue measure and c is an absolute constant. By Fubini's theorem

$$|B| \leq \frac{c}{\lambda} |A|.$$

Similarly we obtain $|C| \leq \frac{c}{\lambda} |B|$ and so $|C| \leq \frac{c^2}{\lambda^2} |A|$. Let us now prove that

$$\{(x,y) \in R^2 : M_2\chi_A(x,y) > 2\lambda\} \subset C$$

Let $(a,b) \notin C$. We want to show $M_2\chi_A(a,b) \leq 2\lambda$. Let $I \times J$ be an open interval of R^2 containing (a,b). We consider

$$S = \{x \in I : (x,b) \in B\}$$

Since (a,b) $\notin$ C we have, by the definition of C, $M^*\chi_{B_1(b)}(a) \leq \lambda$ and so, in particular

$$\frac{|S|_1}{|I|_1} = \frac{|B_1(b) \cap I|_1}{|I|_1} \leq \lambda$$

Also, if $x \in I-S$, i.e. if $x \in I$ and $(x,b) \notin B$, then, by the definition of B, $M^*\chi_{A_2(x)}(b) \leq \lambda$, and so in particular, if T(x) is the set

$$T(x) = \{y \in J : (x,y) \in A\}$$

we have

$$\frac{|T(x)|_1}{|J|_1} = \frac{|A_2(x) \cap J|_1}{|J|_1} \leq \lambda$$

Hence, by Fubini's theorem,

$$|(I \times J) \cap A| \leq \lambda |J|_1 \, |I| + (1-\lambda)|J|_1 \, \lambda |I|_1 < 2\lambda |I \times J|$$

Thus we have for any open set A,

$$|\{M_2\chi_A > 2\lambda\}| \leq |C| \leq \frac{\sigma^2}{\lambda^2} |A|.$$

From this one gets for any measurable set V

$$|\{M_2\chi_V > 2\lambda\}| \leq \frac{c^2}{\lambda^2} |V|$$

and this proves the density property by Theorem 1.2.

Some more applications of the theorems of this section will appear in the following chapters, in particular for the proof that $\mathcal{B}_3$ is not a density basis.

(3) A problem. The result of theorem 1.2. suggests the following question. Assume that the B-F basis $\mathcal{B}$ is invariant only by translations. Is there a similar characterization for the density as the one appearing in 1.2.? Certain results we shall see

in the next sections suggest that a such basis $\mathcal{B}$ is a density basis if and only if there is $r > 0$ such that for each λ, with $0 < \lambda < 1$, there exists a constant $c(\lambda)$ such that for each bounded measurable set A we have

$$|\{M_r \chi_A > \lambda\}| \leq c(\lambda)\ |A|.$$

Here M_r denotes the maximal operator associated to the basis of sets of $\mathcal{B}$ with diameter less than r.

2. INDIVIDUAL DIFFERENTIATION PROPERTIES.

If one knows that a density basis $\mathcal{B}$ differentiates the integral of a function $f \in L$, then one can affirm that the maximal operator associated to $\mathcal{B}$ satisfies a certain weak type property. This is essentially the contents of the main theorem in this section. In order to prove it we shall make use of another important theorem that asserts that the differentiation of integrals of functions by a basis $\mathcal{B}$ is transmitted to smaller functions. This last theorem is due to Hayes and Pauc [1955]. The proof we present here is considerably shorter and simpler. It is based on a idea of Jessen used by Papoulis [1950] for a different purpose, and it is published here for the first time. The main theorem in this section is partly due to Hayes and Pauc [1955], equivalence (a) $\Longleftrightarrow$ (c). The equivalence (a) $\Longleftrightarrow$ (b) and also the proof of the whole theorem appears here for the first time.

THEOREM 2.1. Let $\mathcal{B}$ be a density basis that differentiates $\int f$ for a fixed $f \in L^1(R^n)$, $f \geq 0$. Assume that g is any measurable function such that $0 \leq |g(x)| \leq f(x)$ at every $x \in R^n$. Then $\mathcal{B}$ differentiates $\int g$.

Proof. For a fixed $N > 0$ define

$$f_N(x) = \begin{cases} f(x) & \text{if } f(x) < N \\ 0 & \text{if } f(x) \geq N \end{cases}$$

and let f^N be such that $f(x) = f^N(x) + f_N(x)$ at each $x \in R^n$. By hypothesis $D(\int f,x) = f(x)$ at almost every $x \in R^n$ and also, since $\mathcal{B}$ is a density basis, by Theorem 1.4., $D(\int f_N,x) = f_N(x)$ at almost every $x \in R^n$. So we get at almost every $x \in R^n$, $D(\int f^N,x) = f^N(x)$.

Let us now define

$$g_*(x) = \begin{cases} g(x) & \text{if } f(x) < N \\ 0 & \text{if } f(x) \geq N \end{cases}$$

and g^* such that $g(x) = g_*(x) + g^*(x)$ at each $x \in R^n$. Then we have $|g_*(x)| < N$ at each $x \in R^n$ and so, again by theorem 1.4., $D(\int g_*,x) = g_*(x)$ almost everywhere. Since $|g^*| \leq f^N$, we have at almost every x for each sequence $\{R_k(x)\} \subset \mathcal{B}(x)$ contracting to x.

$$\left|\frac{1}{|R_k(x)|}\int_{R_k(x)} g^*\right| \leq \frac{1}{|R_k(x)|}\int_{R_k(x)} f^N$$

For each $\alpha > 0$ we can then write

$$|\{x : |\overline{D}(\int g,x)-g(x)| > \alpha\}|_e = |\{x : |\overline{D}(\int g^*,x)-g^*(x)|>\alpha\}|_e \leq$$

$$\leq |\{x : |\overline{D}(\int g^*,x)| > \tfrac{\alpha}{2}\}|_e + |\{x : |g^*(x)| > \tfrac{\alpha}{2}\}| \leq$$

$$\leq |\{x : |\overline{D}(\int f^N,x) > \tfrac{\alpha}{2}\}|_e + |\{x : f^N(x) > \tfrac{\alpha}{2}\}| =$$

$$= 2\,|\{x : f^N(x) > \tfrac{\alpha}{2}\}| \leq \frac{4}{\alpha}\int_{f\geq N} f \to 0 \text{ as } N \to \infty.$$

with this we have $\overline{D}(\int g,x) = g(x)$ almost everywhere. Analogously $\underline{D}(\int g,x) = g(x)$ almost everywhere and this proves the theorem.

The following theorem characterizes the derivation by $\mathcal{B}$ of the integral of a function f in terms which are similar to those of the density theorem of Busemann and Feller. It is valid for a general basis.

THEOREM 2.2. Let $\mathcal{B}$ be a basis with the density property. Let $f \geq 0$, $f \in L^1(R^n)$. Then the following three conditions are equivalent:

(a) $\mathcal{B}$ differentiates $\int f$.

(b) For each $\lambda > 0$, each sequence $\{f_k\}$, with $f_k \in L^1$, $f_k \leq f$, $f_k(x) \downarrow 0$ at almost each $x \in R^n$ and for each numerical sequence $\{r_k\}$ with $r_k \downarrow 0$, we have

$$|\{M_k f_k > \lambda\}|_e \to 0$$

where M_k is the maximal operator associated to the basis $\mathcal{B}_{r_k}$ obtained by taking from $\mathcal{B}$ just the elements with diameter less than r_k.

(c) For each $\lambda > 0$, each nonincreasing sequence of measurable sets $\{A_k\}$ with $|A_k| \to 0$ and each numerical sequence $\{r_k\}$ with $r_k \downarrow 0$, we have

$$|\{M_k(f\chi_{A_k}) > \lambda\}|_e \to 0$$

Proof. In order to prove that (a) implies (b) we take an arbitrary open cubic interval Q and $\varepsilon > 0$. We have $f_k \downarrow 0$ pointwise on Q and so, by Egorov's theorem, there is a measurable set A, with $|A| < \varepsilon$, such that $f_k \downarrow 0$ uniformly on $Q - A$. Hence, given $\lambda > 0$, there exists a positive integer h such that $f_k(x) < \lambda$ if $k \geq h$ and $x \in Q - A$. Since we are assuming that $\mathcal{B}$ differentiates $\int f$, by the preceding theorem, $\mathcal{B}$ differentiates $\int f_k$ for each k. Hence, for each $x \in Q - A$ and for each sequence $\{R_j(x)\} \subset \mathcal{B}(x)$ contracting to x we have, as $j \to \infty$

$$\frac{1}{|R_j(x)|} \int_{R_j(x)} f_h \to f_h(x) < \lambda.$$

Therefore

$$\lim_{k\to\infty} \{x \in Q : M_k f_h(x) > \lambda\} \subset A$$

It is clear that, if $k \geq h$, since $f_k \leq f_h$, we have

$$\{x \in Q : M_k f_k(x) > \lambda\} \subset \{x \in Q : M_k f_h(x) > \lambda\}$$

and so

$$\lim_{k\to\infty} |\{x \in Q : M_k f_k(x) > \lambda\}|_e \leq |A| \leq \varepsilon$$

Since Q and ε are arbitrary, we get (b).

That (b) implies (c) is trivial by taking $f_k = f\chi_{A_k}$.

In order to prove that (c) implies (a), let

$$A_k = \{f \geq k\} \text{ for } k = 1, 2, \ldots$$

Since $f \in L^1(R^n)$, $|A_k| \downarrow 0$.

We have, calling $f\chi_{A_k} = f^k$, $f\chi_{A_k'} = f_k$, that $f = f_k + f^k$. Since $\mathcal{B}$ is assumed to be a density basis, for almost every x, $D(\int f_k,x) = f_k(x)$. So for each $\lambda > 0$,

$$|\{x : |\overline{D}(\int f,x) - f(x)| > 2\lambda\}|_e = |\{x : |\overline{D}(f^k,x) - f^k(x)| > 2\}|_e \leq$$

$$\leq |\{x : \overline{D}(\int f^k,x) > \lambda\}|_e + |\{x : f^k(x) > \lambda\}| \leq$$

$$\leq |\{x : M(f\chi_{A_k})(x) > \lambda\}|_e + |\{x : f^k(x) > \lambda\}|$$

The first term in the last member of this chain of inequalities tends to zero by hypothesis as $k \to \infty$. The second one because $f \in L^1(R^n)$. So we get $\overline{D}(\int f,x) = f(x)$ almost everywhere. Similarly $\underline{D}(\int f,x) = f(x)$ almost everywhere. This concludes the proof of the theorem.

REMARKS.

(1) A question of Saks answered by Besicovitch.

Saks [1934] posed the following question. Let $f \in L^1(R^n)$ and assume that $\mathcal{B}_2$ does not differentiate $\int f$. Can it happen

$$\infty > \overline{D}\left(\int f,x\right) > \underline{D}\left(\int f,x\right) > -\infty$$

on a set with positive measure?

The question can of course be formulated for a general basis. Besicovitch [1935] found that the answer for $\mathcal{B}_2$ is negative. His proof will be shown in Chapter IV, together with another partial result in the same direction for more general bases obtained by Guzmán and Menárguez.

(2) A problem regarding Theorem 2.2.

The theorem 1.2. on the density property of a basis that is invariant by homothecies suggests the following question: If one assumes $\mathcal{B}$ in Theorem 2.2. to be invariant by homothecies or by translations, can one simplify conditions (b) and (c) in some similar way as in Theorem 1.2.?

3. DIFFERENTIATION PROPERTIES FOR CLASSES OF FUNCTIONS.

We have already seen in several occasions (see for instance Remark (3) of II.1.) how it is possible to deduce a differentiation property for a class of functions when one knows a weak type inequality for the maximal operator associated to the basis $\mathcal{B}$ one considers. We shall now see how, under certain conditions, one can proceed in the converse direction, obtaining weak type inequalities from differentiation properties.

As in section 1 the initial results, formulated in different terms, were obtained by Busemann and Feller [1934]. The maximal operator was introduced in this context by Guzmán and Welland [1971] and more recently Rubio [1972] and Peral [1974] extended the scope of the initial theorems.

The first theorem, which appears partially in Rubio [1972] characterizes the bases that differentiate $L^1(R^n)$. The proof we offer here is quite trivial once we have Theorem 2.2. at our disposal.

THEOREM 3.1. *Let $\mathcal{B}$ be a differentiation basis in R^n. Then the three following conditions are equivalent:*

(a) *$\mathcal{B}$ differentiates $L^1(R^n)$.*

(b) *For each $\lambda > 0$, each nonincreasing sequence $\{f_k\}$ of functions in $L^1(R^n)$ such that $||f_k||_1 \to 0$, and each numerical sequence $\{r_k\}$ with $r_k \downarrow 0$, one has*

$$|\{M_k f_k > \lambda\}|_e \to 0$$

(c) *For each $\lambda > 0$, each $f \in L^1(R^n)$, each nonincreasing sequence of measurable sets $\{A_k\}$ such that $|A_k| \to 0$, and each numerical sequence $\{r_k\}$ with $r_k \downarrow 0$, one has*

$$|\{M_k(f\chi_{A_k}) > \lambda\}|_e \to 0 \ .$$

Proof. If any of the three conditions (a), (b), (c) holds, then $\mathcal{B}$ is a density basis, by Theorem 1.1. The theorem is then a direct consequence of Theorem 2.2.

When one assumes, as in 1.2., that $\mathcal{B}$ is a B-F basis that is homothecy invariant, one can obtain a simpler expression of the condition (b) in the previous theorem. Then $\mathcal{B}$ differentiates $L^1(R^n)$ if and only if the maximal operator M associated to $\mathcal{B}$ is of weak type (1,1), i.e. if for each $\lambda > 0$ and each $f \in L^1(R^n)$ one has

$$|\{Mf > \lambda\}| \leq c \int \left|\frac{f}{\lambda}\right|$$

c being a constant independent of f and λ. This is essentially the contents of a theorem of Busemann and Feller [1934].

We shall obtain here this theorem as a particular case of another one relative to a basis that is invariant by translations. If $\mathcal{B}$ is only invariant by translations one cannot expect that the differentiability of L^1 imply any weak type condition for the maximal operator. The basis $\mathcal{B}$ obtained by taking all open disks in R^2 with radius less than 1 and all open rectangles such that the length of the smaller side is bigger than 1 is clearly invariant by translations and differentiates L^1. Its maximal operator is not of any weak type. (For details we refer to Peral [1974]). For this reason it is of interest the following result due partly to Rubio [1972] and extended by Peral [1974].

THEOREM 3.2. *Let $\mathcal{B}$ be a B-F basis that is invariant by translations. Then the two following conditions are equivalent*:

(a) *$\mathcal{B}$ differentiates L^1.*

(b) *There exist two constants $c, r > 0$ such that for each $f \in L^1$ and each $\lambda > 0$ one has*

$$|\{M_r f > \lambda\}| \leq c \int \left|\frac{f}{\lambda}\right|$$

where

$$M_r f(x) = \sup \{\frac{1}{|R|} \int_R |f| \;:\; R \in \mathcal{B}(x),\ \delta(R) < r\}$$

In the proof of this theorem we shall make use fo the following lemma due to A.P. Calderón. Its proof can be seen in Zygmund [1959] (vol. II p. 165).

LEMMA 3.3. *Let $\{A_k\}$ be a sequence of measurable sets contained in a fixed cubic interval $Q \subset R^n$ and such that $\sum |A_k| = \infty$. Then there is a sequence $\{x_k\}$ of points in R^n and a set S with positive measure contained in Q such that each $s \in S$ is in*

<u>infinitely many sets of the form</u> $x_k + A_k$.

<u>Proof of Theorem</u> 3.2. That (b) implies (a) is a simple consequence of Theorem 3.1. In order to prove that (a) implies (b) let us prove first that (a) implies the following:

(b*) <u>For each fixed cubic interval</u> Q <u>there exist positive constants</u> $c = c(Q)$, $r = r(Q)$ <u>such that for each non negative</u> $f \in L^1$ <u>with support in</u> Q <u>and each</u> $\lambda > 0$ <u>we have</u>

$$|\{x \in R^n : M_r f(x) > \lambda\}| \leq c \int \frac{f}{\lambda}$$

Assume that (b*) does not hold. Then there is a fixed Q such that for each pair of constants c_k, $r_k > 0$ there is a nonnegative $f_k \in L^1$ supported in Q and also a $\lambda_k > 0$ such that the set

$$E_k = \{x \in R^n : M_{r_k} f_k(x) > \lambda_k\}$$

Satisfies $|E_k| > c_k \int \frac{f_k}{\lambda_k}$.

Let us take a sequence $\{r_k\}$, $r_k \to 0$, such that all numbers r_k are less than the side-length of Q, and let $c_k = 2^k$. We call $g_k = f_k/\lambda_k$ and Q* the cubic interval with the same center as that of Q and three times its size. Clearly $E_k \subset Q^*$ and

$$|E_k| = |\{M_{r_k} g_k > 1\}| > 2^k \int g_k.$$

We can choose for each k a positive integer h_k such that $|Q^*| \leq h_k |E_k| \leq 2|Q^*|$. So we have

$$\sum_{k=1}^{\infty} h_k \; |E_k| = \infty.$$

We consider the sequence $\{A_h\}$ of sets contained in Q* obtained by repeating

h_k times each E_k, i.e. the following sequence:

$$E_1^1,\ E_1^2,\ \ldots,\ E_1^{h_1},\ E_2^1,\ E_2^2,\ \ldots,\ E_2^{h_2},\ E_3^1,\ E_3^2,\ \ldots,\ E_3^{h_3},\ E_4^1,\ \ldots$$

where $E_k^j = E_k$ for each j with $1 \leq j \leq h_k$. Since

$$\sum_{k=1}^{\infty} h_k\ |E_k| = \sum_{h=1}^{\infty} |A_h| = \infty$$

and all sets are contained in Q* we can apply Lemma 3.3. We thus obtain the points

$$x_1^1,\ x_1^2,\ \ldots,\ x_1^{h_1},\ x_2^1,\ x_2^2,\ \ldots,\ x_2^{h_2},\ x_3^1,\ x_3^2,\ \ldots,\ x_3^{h_3},\ x_4^1,\ \ldots$$

and a set S with positive measure contained in Q* such that each point of S is in infinitely many of the sets E_k^j. We define the functions, for each k and each j = 1, 2, ..., h_k,

$$g_k^j(x) = g_k(x - x_k^j)$$

and finally the function

$$f(x) = \sum_{k=1}^{\infty} \alpha_k \sum_{j=1}^{h_k} g_k^j(x)$$

where $\alpha_k > 0$ will be chosen in a moment.

We have

$$||f||_1 \leq \sum_{k=1}^{\infty} \alpha_k h_k\ ||g_k||_1$$

and, since $h_k|E_k| \leq 2|Q^*|$ and $|E_k| > 2^k||g_k||_1$, we get

$$||f||_1 \leq \sum_{k=1}^{\infty} \alpha_k h_k \frac{|E_k|}{2^k} \leq \Big(\sum_{k=1}^{\infty} \frac{2\alpha_k}{2^k}\Big)\ |Q^*|.$$

Let $R \in \mathcal{B}$. We can obviously write

$$\frac{1}{|R|}\int_R f = \sum_{k=1}^{\infty} \alpha_k \sum_{j=1}^{h_k} \frac{1}{|R|}\int_R g_k^j$$

Each $s \in S$ belongs to an infinite number of sets of the form E_k^j. Let these sets be

$$E_{k_1}^{j_{k_1}},\ E_{k_2}^{j_{k_2}},\ \ldots$$

By the definition of the sets E_k^j, there is then a sequence $\{R_h\} \subset \mathcal{B}(s)$, with $R_h \to s$, such that, because of the above equality,

$$\frac{1}{|R_h|}\int_{R_h} f > \alpha_{k_m} \quad \text{for} \quad m = 1, 2, \ldots$$

Let us choose now α_k so that $\alpha_k \to \infty$ and at the same time

$$\sum_{k=1}^{\infty} \frac{\alpha_k}{2^k} < \infty.$$

For example, let us set $\alpha_k = 2^{k/2}$. Then we obtain $f \in L^1$ and at each $s \in S$ we have $\overline{D}(\int f, s) = +\infty$. This contradicts (a). Hence (a) implies (b*).

We have now to deduce (b) from (b*). First of all it is clear, by the invariance by translations of $\mathcal{B}$, that the constants $c(Q)$, $r(Q)$ of (b*) do not depend on the place in R^n where Q is located.

It is also clear that $M_{r/2}f \leq M_r f$ and so we can assume in (b*) that $r(Q)$ is less than half the length of the side of Q. Assume now that $f \geq 0$ is a function in L^1 with support contained in infinitely many disjoint cubic intervals $\{Q_j\}_{j=1}$ each one of them equal in size to Q and such that the distance between any two of them is at least equal to the side length of Q. Then, if r is, as we have assumed, less than half the side length of Q, we clearly have

$$\{M_r f > \lambda\} = \bigcup_{j=1}^{\infty} \{M_r(f\chi_{Q_j}) > \lambda\} = \bigcup_{j=1}^{\infty} H_j.$$

and the sets H_j are disjoint. Hence

$$|\{M_r f > \lambda\}| = \sum_{j=1}^{\infty} |\{M_r(f\chi_{Q_j}) > \lambda\}| \leq \sum_{j=1}^{\infty} c\int_{Q_j} \frac{f}{\lambda} = c\int \frac{f}{\lambda}.$$

Now, for an arbitrary $f \in L^1$, $f \geq 0$, we can set $f = \sum_{h=1}^{\alpha(n)} f_h$ where each f_h is of the type already treated, the functions f_h have disjoint supports and $\alpha(n)$ depends only on the dimension. Thus

$$|\{M_r f > \lambda\}| \leq \sum_{h=1}^{\alpha(n)} |\{M_r f_h > \frac{\lambda}{\alpha(n)}\}| \leq \sum_{h=1}^{\alpha(n)} \alpha(n)\ c\int \frac{f_h}{\lambda} = \alpha(n)\ c\int \frac{f}{\lambda}.$$

The restriction $f \geq 0$ is trivially removed and so we obtain the theorem.

The theorem of Busemann-Feller for a basis that is homothecy invariant is now an easy corollary of Theorem 3.2.

THEOREM 3.4. *Let $\mathcal{B}$ be a B - F basis that is homothecy invariant. Then the two following conditions are equivalent:*

(a) *$\mathcal{B}$ differentiates L^1.*

(b) *The maximal operator M of $\mathcal{B}$ is of weak type (1,1) i.e. there exists a constant $c > 0$ such that for each $f \in L^1$ and each $\lambda > 0$ one has*

$$|\{Mf > \lambda\}| \leq c\int \left|\frac{f}{\lambda}\right|$$

Proof. It is sufficient to prove that for the homothecy invariant basis $\mathcal{B}$ condition (b) of Theorem 3.2. implies condition (b) of this theorem.

Therefore, we assume that there exist $c > 0$ and $r > 0$ such that for each $f \in L^1$ and $\lambda > 0$ we have

$$|\{M_r f > \lambda\}| \leq c\int \left|\frac{f}{\lambda}\right|$$

Take a number $\rho > 0$ and a function $\phi \in L^1$. Define a new function f by setting, for $x \in R^n$,

$$f(x) = \phi(\frac{\rho}{r} x)$$

Observe that

$$c \int |\frac{f(x)}{\lambda}| dx = c \int |\frac{\phi(\frac{\rho}{r} x)}{\lambda}| \, dx = c(\frac{r}{\rho})^n \int |\frac{\phi(z)}{\lambda}| \, dz.$$

In particular, of course, $f \in L^1$. If $y \in R^n$ and $R \in \mathcal{B}_\rho(y)$, we can set

$$\frac{1}{|R|} \int_R |\phi(z)| dz = \frac{1}{|\frac{r}{\rho} R| (\frac{\rho}{r})^n} \int_R |f(\frac{r}{\rho} z)| \, dz =$$

$$= \frac{1}{|\frac{r}{\rho} R| (\frac{\rho}{r})^n} \int_{\frac{r}{\rho}R} |f(x)| (\frac{\rho}{r})^n \, dx = \frac{1}{|\frac{r}{\rho} R|} \int_{\frac{r}{\rho}R} |f(x)| \, dx$$

This prove that $M_\rho \phi(y) = M_r f(\frac{r}{\rho} y)$, since $\frac{r}{\rho} R \in \mathcal{B}_r(\frac{r}{\rho} y)$ and when R runs over all $\mathcal{B}_\rho(y)$, the set $\frac{r}{\rho} R$ runs over all $\mathcal{B}_r(\frac{r}{\rho} y)$. Thus

$$|\{y \in R^n : M_\rho \phi(y) > \lambda\}| = |\{y \in R^n : M_r f(\frac{r}{\rho} y) > \lambda\}| =$$

$$= |\{\frac{\rho}{r} z : M_r f(z) > \lambda\}| = (\frac{\rho}{r})^n \, |\{z : M_r f(z) > \lambda\}| \leq \text{(by hypothesis)} \leq$$

$$\leq (\frac{\rho}{r})^n \, c \int |\frac{f}{\lambda}| = c \int |\frac{\phi}{\lambda}|.$$

This proves that for any $\rho > 0$ and any $\phi \in L^1$ we get

$$|\{M_\rho \, \phi > \lambda\}| \leq c \int |\frac{\phi}{\lambda}|$$

with the same constant. Hence for each $\phi \in L^1$,

$$|\{M\phi > \lambda\}| \leqslant c \int \left|\frac{\phi}{\lambda}\right|$$

and this concludes the proof of the theorem.

REMARKS.

(1) A general context for the theorems of Section 3.

As we shall see in Chapter VI, there are differentiation bases that differentiate, for example, L^5 but not L^3. This motivates the consideration of the following more general situation. Let $\phi : [0,\infty] \to [0,\infty]$ a strictly increasing function, with $\phi(0) = 0$ and such that $\phi(u)$ is of order greater than or equal to the order of u when $u \to \infty$. Let $\phi(L)$ be the collection of all measurable function $f : R^n \to R$ such that $\int \phi(|f|) < \infty$. Assume that a basis $\mathcal{B}$ differentiates $\phi(L)$. What can be said about the maximal operator M of $\mathcal{B}$? The theorems in the following remarks generalize the theorems of section 3 to this situation and are due to Rubio [1972] and Peral [1974].

(2) An extension of Theorem 3.1.

I. Peral [1974] has proved the following result:

Let ϕ be as in Remark (1). Let $\mathcal{B}$ be a B - F basis that differentiates $\phi(L)$. Then for each $\lambda > 0$, for each nonincreasing sequence of real numbers $\{r_k\}$ with $r_k \to 0$ and for each nonincreasing sequence $\{f_k\} \subset \phi(L)$ of nonnegative functions such that

$$\int \phi(f_k) \to 0$$

we have

$$|\{M_{r_k} f_k > \lambda\}| \to 0.$$

The proof does not essentially differ from that of Theorem 3.1.

(3) An extension of Theorem 3.2.

It has been shown also by Peral [1974] that Theorem 3.2. can be generalized to the space $\phi(L)$ in the following way:

Let $\mathcal{B}$ be a translation invariant B - F basis. Assume that $\mathcal{B}$ differentiates $\phi(L)$.

Then there exists a constant $r > 0$, $c > 0$, such that for each $\lambda > 0$ and for each $f \in \phi(L)$ one has

$$|\{M_r f > \lambda\}| \leqslant c \int \phi\left(\frac{2^{n+1}|f|}{\lambda}\right)$$

The reader is invited to prove this theorem by the tecniques already introduced in this chapter and the one used in the next remark.

(4) An extension of Theorem 3.4. A different proof.

B. Rubio [1972] proved the following extension of Theorem 3.4.:

Let $\mathcal{B}$ be a homothecy invariant B - F basis. Assume that $\mathcal{B}$ differentiates $\phi(L)$. Then there exists a constant $c > 0$ such that for each $\lambda > 0$ and each $f \in \phi(L)$, $f \geqslant 0$ one has

$$|\{Mf > \lambda\}| \leqslant c \int \phi\left(\frac{f}{\lambda}\right)$$

The proof given by him is interesting and will be sketched here.

Assume that the theorem is not true. Then, for each $c_k > 0$ there exist $f_k \in \phi(L)$, $f_k \geqslant 0$ and $\lambda_k > 0$ such that

$$|\{Mf_k > \lambda_k\}| > c_k \int \phi\left(\frac{f_k}{\lambda_k}\right)$$

Let us call $g_k = \frac{f_k}{\lambda_k}$. We take a sequence $\{c_k\}$, $c_k > 0$ such that

$$\sum c_k^{-1} < \frac{\phi(1)}{2}.$$

There exists a sequence $\{r_k\}$, $r_k \to 0$, $r_k > 0$, such that

$$|\{M_{r_k} g_k > 1\}| > c_k \int \phi(g_k)$$

There is also a compact subset E_k of $\{M_{r_k} g_k > 1\}$ such that $|E_k| > c_k \int \phi(g_k)$. We consider the open unit cube Q and a fixed k. By using Lemma 1.3. we cover almost completely Q by means of a disjoint sequence $\{E_k^h\}_{h\geq 1}$ of sets homothetic to E_k contained in Q and of diameter less than 1/k. Let ρ_k^h be the homothecy that carries E_k onto E_k^h. We define the function g_k^h by setting

$$g_k^h(\rho_k^h x) = g_k(x)$$

Define then

$$S_k = \sum_{h>1} g_k^h$$

and finally $f = \sup_k S_k$. One easily gets $f \in \phi(L)$ and

$$\int \phi(f) < \frac{1}{2} \phi(1).$$

However, $\overline{D}(\int f,x) \geq 1$ at almost each $x \in Q$. Since $\mathcal{B}$ differentiates $\phi(L)$, we get $f(x) \geq$ ≥ 1 at almost each $x \in Q$ and so

$$\frac{1}{2} \phi(1) > \int \phi(f) \geq \phi(1).$$

This contradiction proves the theorem.

CHAPTER IV

THE INTERVAL BASIS $\mathcal{B}_2$

In the preceding chapters we have already seen several important properties of the interval basis $\mathcal{B}_2$ of R^n. We have proved in II.3 that the maximal operator M_2 associated to $\mathcal{B}_2$ satisfies the following weak type inequality: There exists a constant $c > 0$ such that for each $f \in L_{loc}(R^n)$ and for each $\lambda > 0$ one has

$$|\{M_2 f > \lambda\}| \leqslant c \int \frac{|f|}{\lambda} (1 + \log^+ |\frac{f}{\lambda}|)^{n-1}$$

From this theorem one easyly obtains, as we have indicated there, the following differentiation property:

The basis $\mathcal{B}_2$ of R^n differentiates $L(1 + \log^+ L)^{n-1}(R^n)$. This is the theorem of Jessen-Marcinkiewicz-Zygmund [1935]. We have obtaimed in II.3 a generalization of this result due to Zygmund [1967].

In this chapter we prove several important properties of $\mathcal{B}_2$ and we present also some problems that show more clearly the scope of the general theory developped in Chapter II and III.

1. THE INTERVAL BASIS $\mathcal{B}_2$ DOES NOT SATISFY THE VITALI PROPERTY.

In Caratheodory [1927] appears for the first time a counterexample due to H. Bohr proving that the interval basis does not satisfy the Vitali covering property enjoyed by the cubic intervals. The following is a simplification of that counterexample and is published here for the first time.

THEOREM 1.1. Let Q be the open unit cube in R^2. There exists a subset F of Q, with $|F| = 1$, such that one can assign to each $x \in F$ a sequence $\{I_k(x)\}$ of open unit intervals containing x and contracting to x so that for each disjoint sequence

$\{R_k\}$ extracted from $(I_k(x))_{x\in F,\ k=1,2,\ldots}$ one has

$$|F - \bigcup R_k| > \tfrac{1}{2}.$$

In order to prove the theorem we start with an auxiliary contruction that will also be useful later on.

Auxiliary construction. Let H be an integer bigger than 1 and consider in R^2 the collection of open intervals $I_1, I_2, \ldots, I_H$, obtained as indicated in Figure 4 (where H = 3).

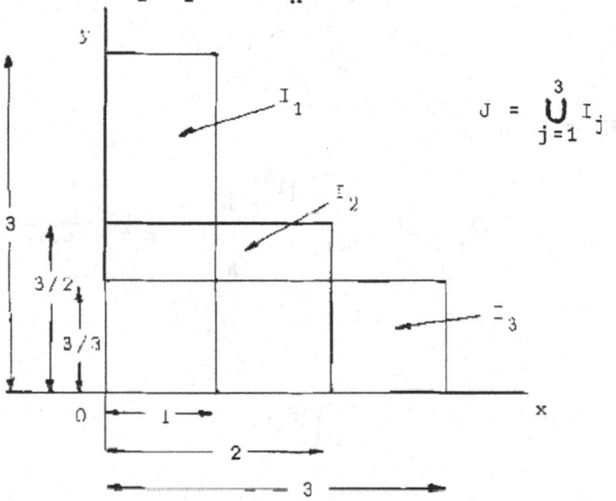

Fig. 4

Each I_j is an open interval with a vertex at 0, a side on the positive part of Ox with length j, and another side on the positive part of Oy with length $\frac{H}{j}$. Hence the area of I_j is H, that of the intersection $E = \bigcap_{j=1}^{H} I_j$ is 1 and that of the union

$$J_H = \bigcup_{j=1}^{H} I_j \quad \text{is} \quad H(1 + \tfrac{1}{2} + \tfrac{1}{3} + \ldots + \tfrac{1}{H}) = H\alpha(H).$$

Proof of Theorem 1.1. Let us consider an increasing sequence $\{H_k\}$ of natural numbers. Fix one of them H_i. By using the lemma III. 1.3. We can cover almost completely the open unit cube Q with a disjoint sequence $\{S_k^i\}_{k=1,2,\ldots}$ of sets homothetic to the J_{H_i} of the auxiliary construction, with all sets S_k^i, k = 1, 2, ... contained in Q and with diameter less than $1/2^i$.

Let $\{I^i_{k,j}\}_{j=1,2,\dots,H_i}$ be the H_i open intervals constituting S^i_k homologous to the intervals I_j of J_{H_i}. Let A^i be the family of all intervals $\{I^i_{k,j}\}_{k\geq 1, j=1,2,\dots,H_i}$. The union of the intervals of A^i is F^i, with $|F^i| = |Q| = 1$. Observe that if $\{R^i_h\} = \{I^i_{k_h,j_h}\}$ is any disjoint sequence of intervals of A_i, then, since we have $I^i_{k,j} \cap I^i_{k,m} \neq \phi$ for $1 \leq j \leq m \leq H_i$, $\{R^i_h\}$ can contain at most one interval $I^i_{k,j}$ of those constituting S^i_k. Furthermore, since for each k and each j = 1, 2, ..., H_i we have

$$\frac{|I^i_{k,j}|}{|S^i_k|} = \frac{1}{\alpha(H_i)}$$

and also $\sum_k |S^i_k| = 1$, we get

$$\sum_{h=1}^{\infty} |R^i_h| = \sum_{h=1}^{\infty} \frac{|I^i_{k_h,j_h}|}{|S^i_{k_h}|} |S^i_{k_h}| \leq \frac{1}{\alpha(H_i)}$$

Let now be

$$F = \bigcap_{i=1}^{\infty} F_i.$$

We have $|F| = 1$. We consider the family

$$A = \bigcup_{i=1}^{\infty} A_i$$

of intervals. To each point x e F we assing the intervals $I_k(x)$ of A containing x. If we now extract from A a disjoint sequence $\{R_k\}$, then we clearly have

$$\sum_{k=1}^{\infty} |R_k| \leq \sum_{k=1}^{\infty} \frac{1}{\alpha(H_k)}$$

Since

$$\alpha(H_k) = (1 + \frac{1}{2} + \dots + \frac{1}{H_k}) > \frac{1}{2} \log H_k,$$

we can choose $\{H_i\}$ so that

$$\sum_{i=1}^{\infty} \frac{1}{\alpha(H_i)} < \frac{1}{2},$$

and so $|F - \bigcup R_k| > \frac{1}{2}$. This concludes the proof of the theorem.

REMARKS.

(1) The basis $\mathcal{B}_2$ does not differentiate L^1.

The auxiliary construction leads inmediately to this result, with the help of the theorem 3.4. of Chapter III. This, in turn, proves that $\mathcal{B}_2$ cannot satisfy the Vitali property, since we know that this implies differentiability of L^1 (cf. I. 3. Remark 9.).

From the auxiliary construction we obtain by looking at the definition of the sets E and J_H,

$$\{M_2\chi_E > \frac{1}{2H}\} \supset J_H$$

since $\frac{|E \cap I_j|}{|I_j|} = \frac{1}{H}$. Hence for each H we get

$$|\{M_2\chi_E > \frac{1}{2H}\}| \gtrsim H\alpha(H)\ |E| \gtrsim\ H \log H\ |E|$$

Therefore M_2 is not of weak type (1,1) and so, by Theorem III. 3.4, $\mathcal{B}_2$ does not differentiate L^1.

In the following section we shall prove something stronger. In a certain sense $\mathcal{B}_2$ differentiates almost no function in L^1.

(2) On the basis $\mathcal{B}_2^*$.

The basis $\mathcal{B}_2$ is regular with respect to $\mathcal{B}_2^*$. Therefore $\mathcal{B}_2^*$ does not satisfy the

Vitali property nor differentiates L^1.

2. SAKS' RARITY THEOREM. A PROBLEM OF ZYGMUND.

The basis $\mathcal{B}_2$ of R^n does not differentiate L but it does differentiate $L(1 + \log^+ L)^{n-1}(R^n)$. Saks [1935] proved a stronger result. For "almost all" functions f of L^1, "almost all" in the sense of Baire's category, one has $\overline{D}(\int f,x) = +\infty$ at each $x \in R^n$. The proof of Saks uses a construction of H. Bohr. Here we shall make use of the auxiliary construction of Section 1. This simplified proof is presented here for the first time.

THEOREM 2.1. The set F of functions f in L^1 such that $\overline{D}(\int f,x) < \infty$ at some point $x \in R^n$ is of the first category in L^1.

Proof. For the sake of clarity, we shall present the proof for n = 2. We show that F is the union of a countable collection of nowhere dense sets in the following way. For k = 1, 2, 3, ... we define F_k as the set of functions f in L^1 such that for some point $x \in R^n$, with $|x| \leq k$, it happens that for all $I \in \mathcal{B}_2(x)$ with $\delta(I) < \frac{1}{k}$ we have $|\int_I f| \leq k|I|$. We clearly have $F = \bigcup_{k=1}^{\infty} F_k$. We now prove that each F_k is nowhere dense, or, what is equivalent that $\overline{F}_k$ has no interior points.

For each k we have $\overline{F}_k = F_k$. In fact, assume that $f_j \to f$ in L^1 and $f_j \in F_k$ for j = 1, 2, ... For each j there is some point x_j, with $|x_j| \leq k$, such that if $I \in \mathcal{B}_2(x_j)$ and $\delta(I) < \frac{1}{k}$, then $|\int_I f_j| \leq k\,|I|$. Since $\overline{B}(0,k)$ is compact, one can extract from $\{x_j\}$ a convergent subsequence. We can assume, changing notation if necessary, that x_j converges to a point x. Clearly, $|x| \leq k$ and if $I \in \mathcal{B}_2(x)$ is such that $\delta(I) < \frac{1}{k}$ we can write

$$|\int_I f| \leq \int_I |f-f_j| + |\int_I f_j|$$

Since I is open, $x \in I$ and $x_j \to x$, we have $I \in \mathcal{B}_2(x_j)$ for j sufficiently large, and so $|\int_I f_j| \leq k|I|$. Furthermore, $\int_I |f-f_j| \to 0$ as $j \to \infty$ and we get $|\int_I f| \leq k|I|$, proving that $f \in F_k$. Therefore $\overline{F}_k = F_k$.

In order to prove that F_k does not contain any interior point we shall use the following lemma which constitutes the kernel of the proof of the theorem. The lemma just means that for each neighborhood V of the origin of L^1 and for each $k = 1, 2, \ldots$ there is a function $\phi_{k,V}$ in that neighborhood which is not in F_k. Its proof is given later.

LEMMA 2.2. For each natural number N there is a nonnegative function ϕ_N such that

(a) $\int \phi_N \leq \frac{1}{N}$

(b) For each $x \in \overline{B}(0,N)$ there exists an interval $I \in \mathcal{B}_2(x)$ with $\delta(I) < \frac{1}{N}$, such that

$$\int_I \phi_N > N\,|I|.$$

With this lemma the fact that F_k does not have any interior points is easily obtained. Let $f \in F_k$. We prove that for each $\eta > 0$ there is a function $g \in L^1 - F_k$ such that $||g - f||_1 \leq \eta$. Let $h \in L^\infty \cap L^1$ be such that $||f - h||_1 \leq \eta/2$. We define $g = h + \phi_N$ where ϕ_N is the function of the lemma with an N that will be chosen in a moment. We can write

$$||g - f||_1 \leq ||h - f||_1 + ||\phi_N|| \leq \frac{\eta}{2} + \frac{1}{N}$$

According to (b) of the lemma, for each $x \in \overline{B}(0,N)$ there exists an interval $I \in \mathcal{B}_2(x)$, with $\delta(I) < \frac{1}{N}$ such that

$$\int_I \phi_N > N|I|$$

We now choose N such that $N \geq k$, $\frac{1}{N} \leq \frac{\eta}{2}$, $N - ||h||_\infty > k$. Then we have $||g - f||_1 \leq \eta$ and

$$\left|\int_I g\right| \geq \left|\int_I \phi_N - \left|\int_I h\right|\right| \geq (N - \|h\|_\infty)|I| > k|I|$$

Hence $g \notin F_k$ as we wanted to prove.

Proof of Lemma 2.2. For the proof of the lemma we start with the simple auxiliary construction of Section 1 with an H that will be conveniently fixed in a moment. By using the lemma III. 1.3. we can cover almost completely the ball $\overline{B}(0,N)$ by means of a disjoint sequence $\{S_k\}$ of sets homothetic to the set J_H of the auxiliary construction contained in $B(0,N)$ and with diameter less than $1/N$. Let $R = \overline{B}(0,N) - \bigcup_1^\infty S_k$. We have $|R| = 0$ and so, for each $\varepsilon > 0$ we can take an open set G containing R and such that $|G| \leq \varepsilon$. For each $x \in R$ we take an open cubic interval $I(x)$ centered at x with diameter less than $1/N$ contained in G. We apply to $(I(x))_{x\in R}$ the theorem of Besicovitch I.1 (cf. Remark 2) obtaining $\{I_k\}$ so that

(a) $R \subset \bigcup I_k \subset G$

(b) $\sum \chi_{I_k} \leq \Theta$

being an absolute constant.

Let us call E_k the set obtained from E of the auxiliary construction by the same homothecy that carries J_H into S_k. We now define the following functions

$$\psi_N(x) = \begin{cases} \dfrac{1}{2N|B(0,N)|} \dfrac{|S_k|}{|E_k|} & \text{if } x \in E_k \\ 0 & \text{if } x \notin E_k \end{cases}$$

$$\rho_k(x) = \begin{cases} 2N & \text{if } x \in I_k \\ 0 & \text{if } x \notin I_k \end{cases}$$

$$\rho_N = \sum_k \rho_k, \qquad \phi_N = \psi_N + \rho_N$$

Then

$$\int \phi_N = \int \Psi_N + \int \rho_N = \frac{1}{2N|B(0,N)|} \sum_{k=1}^{\infty} |S_k| + \sum_{k=1}^{\infty} \int \rho_k =$$

$$= \frac{1}{2N} + \sum_{k=1}^{\infty} \int 2N\chi_{I_k} \leq \frac{1}{2N} + 2\Theta N\varepsilon$$

and so

$$\int \phi_N < \frac{1}{N} \quad \text{if } \varepsilon < \frac{1}{4N^2\Theta}.$$

Now if $x \in \overline{B}(0,N)$ is in some S_k, then it will be in some of the intervals I_k^j, $j = 1, 2, \ldots, H$ composing S_k and I_k^j has diameter less than $1/N$. So we get

$$\int_{I_k^j} \phi_N \geq \frac{1}{2N} \frac{|S_k|}{|E_k|} |E_k| = \frac{1}{2N} \frac{|S_k|}{|I_k^j|} |I_k^j| = \frac{\alpha(H)}{2N} |I_k^j|$$

where $\alpha(H) = 1 + \frac{1}{2} + \ldots + \frac{1}{H}$. If we choose H so that $\frac{1}{2N} \alpha(H) > N$, then

$$\int_{I_k^j} \phi_N > N\, |I_k^j|.$$

If $x \notin S_k$ then $x \in I_k$ for some k and by the definition of ρ_k and ϕ_N,

$$\int_{I_k} \phi_N > N\, |I_k|.$$

This concludes the proof of the lemma.

The theorem we have proved suggests the following problem proposed by A. Zygmund.

Problem 2.3. Given $f \in L^1(R^2)$ is it possible to select a pair of rectangular directions so that if $\mathcal{B}$ is the basis of all open rectangles with sides in those directions then $\mathcal{B}$ differentiates $\int f$? If the answer is affirmative, how is the set of all eligible directions with this property?

An afirmative answer to this problem would be very useful, since for many considerations, given $f \in L^1$ one can freely choose the coordinate axes.

3. A THEOREM OF BESICOVITCH ON THE POSSIBLE VALUES OF THE UPPER AND LOWER DERIVATIVES.

We consider the basis $\mathcal{B}_2$ in R^2. Let $f \in L^1(R^2)$. Since $\mathcal{B}_1 \subset \mathcal{B}_2$ and $\mathcal{B}_1$ differentiates $\int f$, it is easy to see that with respect to the basis $\mathcal{B}_2$ we have

$$\overline{D}(\int f,x) \geq f(x) \geq \underline{D}(\int f,x)$$

at almost every $x \in R^2$. However, according to the previous sections, it can happen that the sets

$$\{x \in R^2 : f(x) < \overline{D}(\int f,x)\}$$

$$\{x \in R^2 : \underline{D}(\int f,x) < f(x)\}$$

have positive measure. With respect to this situation Saks [1934] proposed the following question: <u>Can any of the sets</u>

$$\{x \in R^2 : f(x) < \overline{D}(\int f,x) < \infty\}$$

$$\{x \in R^2 : -\infty < \underline{D}(\int f,x) < f(x)\}$$

<u>be of positive measure?</u> The negative answer is due to Besicovitch [1935]. Here we present the result of Besicovitch. In the remark at the end we show how the theorem can be somewhat extended.

THEOREM 3.1. *We consider the interval basis $\mathcal{B}_2$ in R^2. Let $f \in L^1(R^2)$ be a fixed function. Then the two sets*

$$\{x \in R^2 : f(x) < \overline{D}(\int f, x) < \infty\}$$

$$\{x \in R^2 : -\infty < \underline{D}(\int f, x) < f(x)\}$$

have measure zero.

Proof. We shall carry out the proof for the first of these sets. For the other set one can put $g = -f$ and apply the result for the first set to g.

It will be enough to prove that for α, β, γ rational such that $0 < \alpha < \beta < \gamma$, the set

$$E(\alpha, \beta, \gamma) = \{x \in R^2 : f(x)+\alpha < \overline{D}(\int f, x) < f(x)+\beta,\ -\gamma < f(x) < \gamma\}$$

has measure zero.

Let us assume that there are three numbers α, β, γ as above so that the set $E = E(\alpha, \beta, \gamma)$ is of positive measure. Let us set $f = f_\gamma + f^\gamma$ where

$$f_\gamma(x) = \begin{cases} f(x) & \text{if } |f(x)| \leq \gamma \\ 0 & \text{if } |f(x)| > \gamma \end{cases}$$

Since $\mathcal{B}_2$ differentiates L^∞ and $f_\gamma \in L^\infty$, we have $D(\int f_\gamma, x) = f_\gamma(x)$ almost everywhere. Let us call $\tilde{E}$ the subset of E where $D(\int f_\gamma, x) = f_\gamma(x)$. We have $|\tilde{E}| = |E| > 0$.

It $x \in \tilde{E}$ we have

$$\overline{D}(\int f, x) = \overline{D}(\int (f_\gamma + f^\gamma), x) = \overline{D}(\int f^\gamma, x) + f_\gamma(x)$$

and therefore, having in mind the definition of E, if $x \in \tilde{E} \subset E$ we have

$$f_\gamma(x) + f^\gamma(x) + \alpha < \overline{D}(\int f,x) = \overline{D}(\int f^\gamma,x) + f_\gamma(x) <$$

$$< f_\gamma(x) + f^\gamma(x) + \beta$$

and $-\gamma < f(x) < \gamma$, i.e. $f^\gamma(x) = 0$.

Hence, <u>if</u> $x \in \tilde{E}$ we can write

$$\alpha < \overline{D}(\int f^\gamma,x) < \beta$$

and $f^\gamma(x) = 0$. Thus the set

$$\hat{E} = \{x \in R^2 : \alpha < \overline{D}(\int f^\gamma,x) < \beta,\ f^\gamma(x) = 0\}$$

has positive measure. We now intend to prove that this leads to a contradiction and this will conclude the proof of the theorem.

For each s, with $0 < s < 1$, we define $\hat{E}_s$ as the subset of points x of $\hat{E}$ such that for each $I \in \mathcal{B}_2(x)$, with $\delta(I) < s$, one has

$$\frac{1}{|I|}\int_I f^\gamma < \beta.$$

The set $\hat{E}$ is the union of all $\hat{E}_s$ with $s \in Q$, $s > 0$, and so there must be some $s^* > 0$ such that $\hat{E}_{s^*}$ is of positive exterior measure. Let E^* be a subset of $\hat{E}_{s^*}$ with finite positive exterior measure. Let H be an open set such that $H \supset E^*$, $|H| \leq (1+\eta)|E^*|_e$ with an $\eta > 0$ that will be conveniently chosen later.

For each $x \in E^*$ one can choose $I(x) = I \subset H$, $I = I(x) \in \mathcal{B}_2(x)$, with $\delta(I(x)) <$ $< s^*$ such that

$$\alpha < \frac{1}{|I|}\int f^\gamma < \beta.$$

We shall now apply the following lemma whose proof is presented at the end.

LEMMA 3.2. *Let* I *be any open interval and* $f^\gamma \in L^1$ *such that for each* $x \in I$ *we have either* $f^\gamma(x) = 0$ *or* $|f^\gamma(x)| > \gamma$. *Let us assume that*

$$\frac{1}{|I|}\int_I f^\gamma = \rho \text{ with } 0 < \rho < \gamma.$$

Then one can choose a disjoint sequence $\{I_k\}$ *of open intervals contained in* I *such that for each* k

$$\frac{1}{|I_k|}\int_{I_k} f^\gamma = \gamma \quad \text{and} \quad |\bigcup_k I_k| \geq \frac{\rho}{\gamma}|I|.$$

To each I(x) we apply this lemma, obtaining $\{I_k(x)\}$ satisfying $I_k(x) \subset I(x) \subset \subset H$,

$$|\bigcup_k I_k(x)| > \frac{\alpha}{\gamma}|I(x)|, \; \frac{1}{|I_k(x)|}\int_{I_k(x)} f^\gamma = \gamma$$

for each k.

Let now be $G = \bigcup_{x \in E^*} I(x)$,

$$U = \{I_k(x) : x \in E^*, k = 1, 2, \ldots\}$$

and let $\tilde{E} = G - U$. Clearly $\tilde{E}$ is measurable, since G and U are open. Furthermore $\tilde{E} \supset \supset E^*$. In fact, for each $y \in E^* \subset \hat{E}_{s^*}$ and each interval $I \in \mathcal{B}_2(y)$ with $\delta(I) < s^*$ we have, by the definition of $\hat{E}_{s^*}$,

$$\frac{1}{|I|}\int_I f^\gamma < \beta$$

However, for each one of the sets $I_k(x)$ composing U we have $\delta(I_k(x)) < s^*$ and

$$\frac{1}{|I_k(x)|}\int_{I_k(x)} f^\gamma = \gamma > \beta.$$

So $y \notin U$. This implies $E^* \subset G - U = \tilde{E}$.

Now for each $I(x)$ with $x \in E^*$ we have

$$\frac{|I(x) \cap U|}{|I(x)|} \geq \frac{|\bigcup_k I_k(x)|}{|I(x)|} > \frac{\alpha}{\gamma}$$

and so

$$G \subset \{x \in R^2 : M_2\chi_U(x) > \frac{\alpha}{\gamma}\}$$

Hence, by Theorem III. 1.2. we have $|G| \leq c|U|$. Where c depends only on α/γ. Therefore $|\tilde{E}| \leq c|U|$. We clearly have $U \cup \tilde{E} \subset H$ since each $I(x)$ is in H. So we get

$$(1+\eta)\ |E^*|_e \geq |H| \geq |U| + |\tilde{E}| \geq |\tilde{E}|(1+\frac{1}{c}) \geq |E^*|_e\ (1+\frac{1}{c})$$

This is impossible if we choose $\eta < \frac{1}{c}$. This contradiction concludesthe proof of the theorem.

Proof of Lemma 3.2. We take the interval I and perform on it the following process P. If d is the smaller side and $|d|_1$ its length, D is the bigger side and $|D|_1$ its length, we divide D into equal partes of length between $\frac{|d|_1}{2}$ and $\frac{|d|_1}{4}$. Once we have the dividing points, we draw through them lines parallel to d. We get a partition of I into a certain number of partial intervals $\{I_1^*. I_2^*, \ldots, I_N^*\}$ such that the ratio between their bigger sides and their smaller ones is between 2 and 4. If the mean of f^γ on each one of the partial intervals is less than γ, then the process P on I is finished. If there is some partial interval I_j^* such that the mean of f on I_j^* is bigger than or equal to γ, we take a maximal interval contained in I and containing I_j^* such that the mean of f^γ on it is exactly γ. In this way we have partitioned I into a finite disjoint collection $\mathcal{R}$ of intervals on which the mean of f^γ is

γ and another finite disjoint collection $\mathcal{R}^*$ constituted by the initial partial intervals I_j^* or by those parts of them obtained in the process of construction of the intervals of $\mathcal{R}$. On the intervals of $\mathcal{R}^*$ the mean of f^γ is less than γ. This finishes the process P on I.

We now keep the intervals of $\mathcal{R}$ and on each interval of $\mathcal{R}^*$ we perform the same process P. In this way we get a sequence $\{I_k\}$ of disjoint intervals, with $\bigcup I_k \subset I$ and such that

$$\frac{1}{|I_k|}\int_{I_k} f^\gamma = \gamma$$

On the other hand, if $x \in I - \bigcup I_k$, then there exists a sequence of intervals $H_k(x) \subset \subset \mathcal{B}_2(x)$ contracting to x such that the ratio between the bigger side an the smaller side is between 2 and 4 and such that

$$\frac{1}{|H_k(x)|}\int_{H_k(x)} f^\gamma < \gamma.$$

Since $f \in L$, and the sequence $\{H_k(x)\}$ is regular with respect to squares, we get

$$\frac{1}{|H_k(x)|}\int_{H_k(x)} f^\gamma \to f^\gamma(x)$$

for almost every $x \in I - \bigcup I_k$ and, therefore, $f^\gamma(x) \leq \gamma$ and so $f^\gamma(x) \leq 0$ at almost every $x \in I - \bigcup I_k$. So we can write

$$\rho|I| = \int_I f^\gamma = \int_{I-\bigcup I_k} f^\gamma + \int_{\bigcup I_k} f^\gamma \leq \int_{\bigcup I_k} f^\gamma \leq \gamma|\bigcup I_k|$$

This concludes the proof of the lemma.

REMARKS.

(1) An extension of the theorem of Besicovitch.

The theorem 3.1. can be extended to a B - F basis $\mathcal{B}$ that is invariant by homothecies such that all its sets are convex. One obtains the following result that belongs to the author and M.T. Menárguez and is published here for the first time.

THEOREM. Let $f \in L^1$, $f \geq 0$ and let $\mathcal{B}$ be a B - F basis that is invariant by homothecies and such that all sets of $\mathcal{B}$ are convex and with center of symmetry. Assume that $\mathcal{B}$ is a density basis. Then the two sets

$$\{x \in R^n : f(x) < \overline{D}(\int f,x) < \infty\}$$

$$\{x \in R^n : \underline{D}(\int f,x) < f(x)\}$$

have measure zero.

The proof of the theorem is carried out following the pattern of the one of the theorem of Besicovitch. Only the lemma used there has to be replaced by the following one.

LEMMA. Let R be an open bounded set that is convex and with center of symmetry. Let $f^\gamma \in L^1(R^n)$ be a function such that for each $x \in R^n$ one has either $f^\gamma(x) = 0$ or else $f^\gamma(x) > \gamma > 0$. Assume

$$0 < \alpha < \frac{1}{|R|} \int_R f^\gamma < \gamma$$

Let $\mathcal{R}$ be the differentiation basis such that, for each $x \in R^n$, $\mathcal{R}(x)$ is the collection of all dilatations and contractions of R with respect to its center of symmetry translated to be centered at x.

Then there exists a disjoint sequence $\{R_k\} \subset \mathcal{R}$ of sets contained in R*, the dilatation of R with ratio 3 and center that of R, such that

$$\frac{1}{|R_k|} \int_{R_k} f^\gamma \geq \gamma \quad \text{and} \quad |R_k| > c|R|$$

where c depends on α, γ, and on the dimension n.

Proof. The basis $\mathcal{R}$ differentiates $\int f^{\gamma}$ since it is regular with respect to cubic intervals. Let H be the null set where $D(\int f^{\gamma}\chi_R, x)$ does not exist or it is not $(f^{\gamma}\chi_R)(x)$. Let

$$S = \{x \in R - H : f^{\gamma}(x) > \gamma\}$$

If $x \in S$, then there exists some $R(x) \in \mathcal{R}(x)$ such that

$$\frac{1}{|R(x)|} \int_{R(x)} f^{\gamma}\chi_R = \gamma$$

Let R(x) be the largest element of $\mathcal{R}(x)$ such that this equality holds. Take any $T \in \mathcal{R}(x)$ such that $T \supset R$. Then

$$\frac{1}{|T|} \int_T f^{\gamma}\chi_R \leq \frac{1}{|R|} \int_R f^{\gamma} < \gamma.$$

Hence R(x) cannot contain R and so it must be contained in R* (One can easily see that R* is the union of all sets of $\mathcal{R}$ centered at some point of R and not containing the set R).

Once we have assigned to each $s \in S$ one set R(x) we proceed in the following way. Let $d_1 = \sup\{\delta(R(x)) : x \in S\}$. We take $x_1 \in S$ such that $\delta(R(x_1)) > \frac{3}{4} d_1$. Let $R^*(x_1)$ be the union of all sets T of $\mathcal{R}$ with diameter $2d_1$ such that $T \cap R(x_1) \neq \phi$. We have that $R^*(x_1) \in R$ and $|R^*(x_1)| \leq c(n)\, |R(x_1)|$ where c(n) is a constant depending only on the dimension. Let $d_2 = \sup\{\delta(R(x)) : x \in S - R^*(x)\}$ We take $x_2 \in S - R^*(x_1)$ such that $\delta(R(x_2)) > \frac{3}{4} d_2$ and construct $R^*(x_2)$ as before, etc.... So we obtain $R(x_i) \cap R(x_j) = \phi$ for $i \neq j$ and also $S \subset R^*(x_j)$, $\bigcup R(x_i) \subset R^*$. Therefore

$$\alpha|R^*| \leq 3^n\alpha|R| < 3^n\int_R f^{\gamma} = 3^n\int_S f^{\gamma} \leq 3^n \sum \int_{R^*(x_j)} f^{\gamma}\chi_R \leq$$

$$\leq 3^n\, \gamma \sum_j |R^*(x_j)| = 3^n\, \gamma \sum_j c(n)\, |R(x_j)|.$$

We have used the inequality

$$\int_{R^*(x_j)} f^{\gamma} \chi_R < \gamma |R^*(x_j)|$$

which is obtained by observing that $R^*(x_j)$ is bigger than $R(x_j)$ and this set was supposed to be the biggest one among those of $\mathcal{R}(x)$ such that

$$\int_{R(x_j)} f^{\gamma} \chi_R = \gamma |R(x_j)|$$

So we get

$$\sum_j |R(x_j)| = |\bigcup R(x_j)| > c|\bar{R}|,$$

and the sequence $\{R(x_j)\}$ satisfies all the requirements, of the lemma since we have also.

$$\frac{1}{|R(x_j)|} \int_{R_j(x)} f^{\gamma} \geq \gamma.$$

CHAPTER V

THE BASIS OF RECTANGLES $\mathcal{B}_3$

The basis $\mathcal{B}_3$ of all open rectangles of R^2 raises many interesting problems in differentiation theory. The basis $\mathcal{B}_3$ includes all sets in $\mathcal{B}_2$ and therefore it does not have the Vitali property. More complicated is to prove that $\mathcal{B}_3$ is not a density basis.

Zygmund was the first in pointing out that, as a consequence of a paradoxical set contructed by Nikodym [1927] for other purposes, $\mathcal{B}_3$ does not satisfy the density property. We describe in Section 3 the set of Nikodym and its relation with differentiation. Busemann and Feller [1934] gave a proof of this same fact based on a construction proposed by Besicovitch [1928] for the solution of a problem proposed by Kakeya [1917]. This problem consist in finding out what is the infimum of the areas of those sets in R^2 such that a needle of length 1 can be continuously moved within the set so that at the end it occupies the original place but in inverted position. This problem is also referred to as the "needle problem". Section 1 is devoted to a construction that leads to the solution of such problem and Section 2 deals with the application of that construction to differentiation theory.

1. THE PERRON TREE. THE KAKEYA PROBLEM.

The construction leading to the solution of the Kakeya problem is usually called the Perron tree. Perron [1928] simplified the original construction of Besicovitch [1928]. Later on this construction has been further simplified by Rademacher [1962] and Schoenberg [1962]. The following theorem is essentially the Rademacher constructionwith some slight modifications that will make it more useful for our purposes.

THEOREM 1.1. *Consider in* R^2 *the* 2^n *open triangles* $\{A_h\}_{h=1}^{2^n}$ *obtained by joining the point* $(0,1)$ *with the points* $(0,0)$, $(1,0)$, $(2,0)$, $(3,0)$, ..., $(2^n,0)$. *Let* A_h *be the triangle with vertices* $(0,1)$, $(h-1,0)$, $(h,0)$. *Then, given* α, *with* $\frac{1}{2} < \alpha < 1$, *it* is possible to make a parallel translation of each A_h along the axis Ox to a new position $\overline{A}_h$ so that one has

$$\left|\bigcup_{h=1}^{2^n} \overline{A}_h\right| \leq (\alpha^{2n} + 2(1-\alpha)) \left|\bigcup_{h=1}^{2^n} A_h\right|$$

Proof. The theorem will be obtained by repetition of the following process that, for reference purposes, we shall call the basic construction.

Basic construction. Consider two adjacent triangles T_1, T_2 with basis on Ox, with the same basis length b and with height length h, as in Figure 5. Let $0 < \alpha < 1$. Keeping T_1 fixed we shift T_2 towards T_1 to position T_2^* in such a way that the sides that are not parallel meet at a point at distance αh from Ox as in Figure 6.

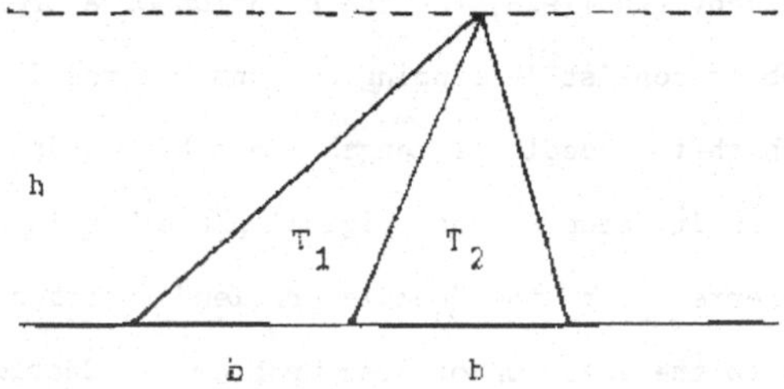

Fig. 5

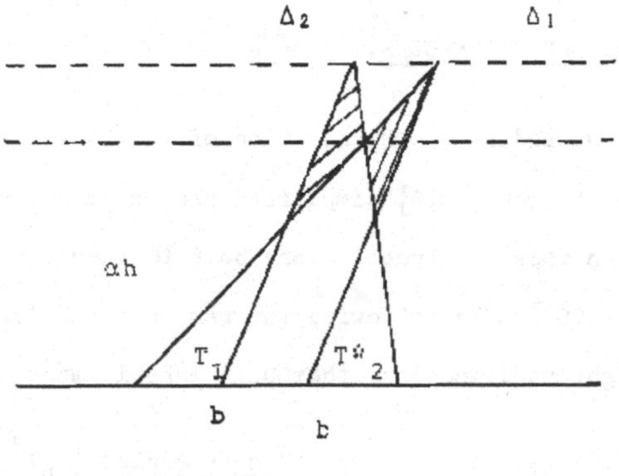

Fig. 6

The union of T_1 and T_2^* is composed by a triangle S (not shaded portion in Fig. 6) homothetic to $T_1 \cup T_2$ plus two "excess triangles" Δ_1, Δ_2. One can easily get

$$|S| = \alpha^2 |T_1 \cup T_2|$$

$$|\Delta_1| + |\Delta_2| = 2(1-\alpha)^2 |T_1 \cup T_2|$$

and so

$$|T_1 \cup T_2^*| = |\alpha^2 + 2(1-\alpha)^2| \; |T_1 \cup T_2|.$$

We shall now apply this basic construction to the situation of the theorem. Consider the 2^{n-1} pairs of adjacent triangles (A_1,A_2), (A_3,A_4), ..., (A_{2^n-1},A_{2^n}). To each pair we apply the basic contruction with the same α given in the statement of the theorem. We obtain the triangles S_1, S_2, ..., $S_{2^{n-1}}$ and the excess triangles Δ_1^1, Δ_2^1; Δ_1^2, Δ_2^2; ..., $\Delta_1^{2^n-1}$, $\Delta_2^{2^n-1}$. We now shift S_2 along Ox towards S_1 so that it becomes $\tilde{S}_2$ adjacent to S_1. Then we shift S_3 to position $\tilde{S}_3$ adjacent to $S_1 \cup \tilde{S}_2$, and so on. In these motions each S_h must carry with it the two excess triangles Δ_1^h, Δ_2^h, so that what we are in fact doing is equivalent to shifting the triangles A_2, A_3, ..., A_{2^n}, to some new positions $\tilde{A}_2$, $\tilde{A}_3$, ..., $\tilde{A}_{2^n}$. Consider now $A_1 \cup \tilde{A}_2 \cup \tilde{A}_3 \cup \dots \cup \tilde{A}_{2^n}$. This figure is composed by 2^{n-1} triangles S_1, $\tilde{S}_2$, ..., $\tilde{S}_{2^{n-1}}$, whose union is of area

$$\alpha^2 |A_1 \cup A_2 \cup A_3 \cup \dots \cup A_{2^n}|,$$

plus shifted excess triangles, whose union is of area not larger than

$$2(1-\alpha)^2 \; |A_1 \cup A_2 \cup A_3 \cup \dots \cup A_{2^n}|$$

The 2^{n-1} triangles S_1, $\tilde{S}_2$, $\tilde{S}_3$, ..., $\tilde{S}_{2^{n-1}}$ are in the same situation as the initial triangles A_1, A_2, A_3, ..., A_{2^n}. One subjects them to the same process, always carrying the excess triangles so that in fact one moves the entire triangles A_2, A_3, ..., A_{2^n}.

This process is repeated n times and at the end one obtains a figure $A_1 \cup \bar{A}_2 \cup \bar{A}_3 \cup \cup \ldots \cup \bar{A}_{2^n}$ which is composed by a triangle H homothetic to $A_1 \cup A_2 \cup \ldots \cup A_{2^n}$ of area

$$\alpha^{2n}|A_1 \cup A_2 \cup \ldots \cup A_{2^n}|$$

plus additional triangles whose union has an area not larger than

$$|\bigcup_{h=1}^{2^n} A_h| \; [2(1-\alpha)^2 + 2\alpha^2(1-\alpha)^2 + 2\alpha^4(1-\alpha)^2 + \ldots + 2\alpha^{2(n-1)}(1-\alpha)^2]$$

Hence, if we set $A_1 = \bar{A}_1$, we get

$$|\bigcup_1^{2^n} \bar{A}_h| \leq (\alpha^{2n}+2(1-\alpha)^2 \frac{1}{1-\alpha^2}) \; |\bigcup_1^{2^n} A_h| < |\alpha^{2^n}+2(1-\alpha)| \; |\bigcup_1^{2^n} A_h|.$$

This concludes the proof of the theorem.

It is clear that one can perform an affine transformation in the situation of Theorem 1.1. in order to give it a more flexible structure. Parallel lines keep being parallel after the transformations and ratios between areas of figures do not change. So one easily arrives to the following result.

THEOREM 1.2. Let A B C be a triangle of area H. Given any $\varepsilon > 0$ it is possible to partition the basis B C into 2^n parts $I_1, I_2, I_3, \ldots, I_{2^n}$ (n depends on ε) and to shift the triangles $T_1, T_2, T_3, \ldots, T_{2^n}$ with basis $I_1, I_2, I_3, \ldots, I_{2^n}$ and common vertex A along B C to positions $\bar{T}_1, \bar{T}_2, \bar{T}_3, \ldots, \bar{T}_{2^n}$ so that

$$|\bar{T}_1 \cup \bar{T}_2 \cup \bar{T}_3 \cup \ldots \cup \bar{T}_n| < \varepsilon H.$$

Proof. We first take α so that $0 < \alpha < 1$, $2(1-\alpha) < \varepsilon/2$, and then take n so that $\alpha^{2n} < \frac{\varepsilon}{2}$. We now consider the result of Theorem 1.1. with this n and α, and an affine transformation ρ that carries (0,1) to A, (0,0) to B and $(2^n,0)$ to C. Then $\rho(A_h) = T_h$ and $\rho(\bar{A}_h) = \bar{T}_h$ for $h = 1, 2, \ldots, 2^n$.

The set $\bigcup_{h=1}^{2^n} \overline{T}_h$ is usually called a Perron tree, because of the ramifications due to the excess triangles. With the preceding result one easily obtains the solution of the needle problem.

THEOREM 1.3. Given $\eta > 0$ and a straight segment A B of length 1 in R^2 one can construct a figure F with area less than η so that one can continuously move that segment within F so that it finally occupies the same place but in inverted position.

Proof. First of all we show that one can continuously move a segment from one straight line to another one parallel to it sweeping out an area as small as one wishes. It is enough to observe in Figure 7 that one can move A B to A_4 B_4 sweeping out the area of the shaded portion which can be made as small as one wishes taking

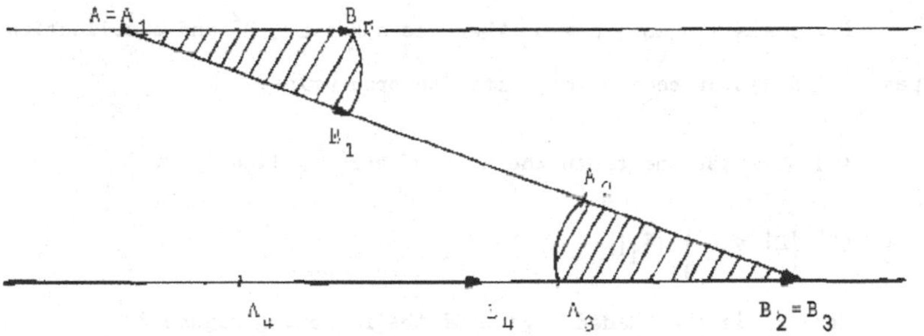

Fig. 7

A B_3 sufficiently large.

We now show that A B can be moved to a straight line forming an angle of 60° with its original position within a figure of area less than $\eta/6$. Six repetitions of the same process will give us the figure F of the theorem. Let M N P be an equilateral triangle of area equal to 10 placed so that A B is in the interior of M N. Observe that the height of M N P is bigger than 1. To M N P we apply Theorem 1.2. taking as basis N P and with an ε such that $10\,\varepsilon < \frac{\eta}{12}$. We obtain the triangles $\overline{T}_1, \overline{T}_2, \ldots, \overline{T}_{2^n}$. The segment A B can be continuously moved within $\overline{T}_1$ from M N to the other side of $\overline{T}_1$ not on N P. From there one can move the segment to the side of $\overline{T}_2$ parallel to it sweeping an area less than $\frac{\eta}{12 \times 2^n}$. Now we move it again within $\overline{T}_2$ to the other side of $\overline{T}_2$ not on N P, and so on. The area swept out in this process is less than $\eta/6$, and the needle is at the end on a line forming an angle of 60° with the original position.

REMARKS.

(1) Applications of the Perron tree.

In the following sections we show some interesting applications connected with differentiation theory. Fefferman [1971] has used it as the basic tool for proving that the characteristic function of the unit disc in R^2 is not a multiplier in the following form:

Fix a small number $\eta > 0$. There is a set $E \subset R^2$ and a collection $\mathcal{R} = \{R_j\}$ of pairwise disjoint rectangles, with the properties:

(1) At least one-tenth the area of each $\tilde{R}_j$ lies in E.

(2) $|E| \leq \eta \sum_j |R_j|$.

Here R_j is the shaded region of the following figure 8

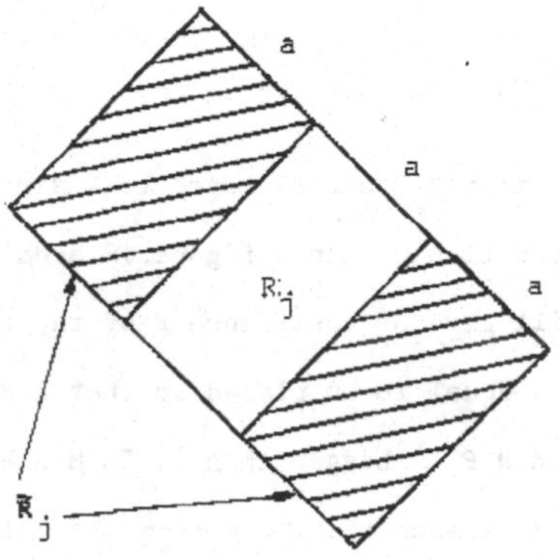

Fig. 8

2. THE BASIS $\mathcal{B}_3$ IS NOT A DENSITY BASIS.

The result of the preceding section allows us to deduce immediately that $\mathcal{B}_3$ is not a density basis. One can even prove something a little stronger. We shall see that there is a B - F basis $\mathcal{B}$ invariant by homothecies generated by a sequence of open rectangles $\{R_k\}$, i.e. $B \in \mathcal{B}$ if and only if B is homothetic to R_k for some k, such that $\mathcal{B}$ is not a density basis.

In order to construct this basis $\mathcal{B}$ we proceed as follows. If A_h is the triangle defined in Theorem 1.1. we call R_h the open rectangle one of whose sides is the segment joining the points (0,1) and (h,0) and such that the side parallel to this one passes through (h-1,0), as indicated in Figure 9.

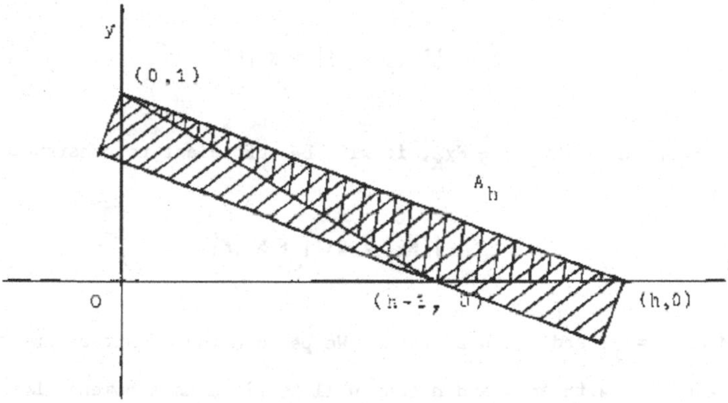

Fig. 9

We now consider the minimal B - F basis $\mathcal{B}$ invariant by homothecies containing all rectangles R_h, h = 1, 2, ... Together with $\mathcal{B}$ we consider also the minimal B - F basis $\bar{\mathcal{B}}$ invariant by homothecies containing all triangles A_h, h = 1, 2, 3, ... If M is the maximal operator associated to $\mathcal{B}$ and $\bar{M}$ that associated to $\bar{\mathcal{B}}$, since we clearly have $|R_h| = 2|A_h|$, for any $f \in L$ and any h we have

$$\frac{1}{|A_h|}\int_{A_h}|f| \leq \frac{2}{|R_h|}\int_{R_h}|f|$$

And so we easily deduce for every $x \in R^2$,

$$\overline{M}f(x) \leq 2M\, f(x)$$

With this one obtains the following result.

THEOREM 2.1. The basis $\mathcal{B}$ described above is not a density basis.

Proof. Since $\mathcal{B}$ is homothecy invariant, we can apply the density criterion of Busemann and Feller given in Theorem III. 1.2. It will be sufficient to prove that there exists a λ, $0 < \lambda < 1$, such that for each positive constant N we can construct a bounded measurable set K so that

$$|\{M\chi_K > \lambda\}| > N\,|K|$$

Observe that, since $M\chi_K \geq \frac{1}{2}\overline{M}\chi_K$, it will be sufficient to construct K so that

$$|\{\overline{M}\chi_K > 2\lambda\}| > N\,|K|.$$

Let us take $\lambda = \frac{1}{16}$ and let N be given. We perform the constructions of the statement of Theorem 1.1., with an α and n that will be fixed in a moment. Let K be the set $\bigcup_1^{2n} \overline{A}_h$. We now show that all points of the open shaded portion V in Figure 10 are in the set $\{\overline{M}\chi_K > \frac{1}{8}\}$.

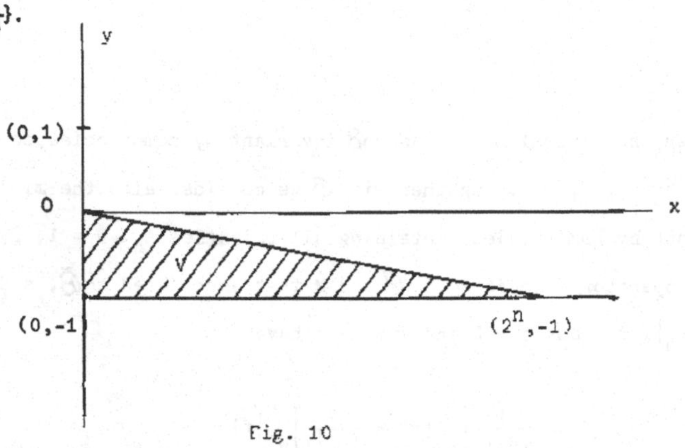

Fig. 10

Once we prove this, we can write

$$|\{M\chi_K > \tfrac{1}{16}\}| \geq |\{\bar{M}\chi_K > \tfrac{1}{8}\}| \geq |V| = |\bigcup_1^{2^n} A_h| \geq (\alpha^{2n}+2(1-\alpha))^{-1}|K|$$

so it is enough to choose α and n so that

$$[\alpha^{2n} + 2(1-\alpha)]^{-1} > N$$

and we get $|\{M\chi_K > \frac{1}{16}\}| > N|K|$ and this will conclude the proof of the theorem.

In order to prove that prove that $V \subset \{\bar{M}\chi_K > \frac{1}{8}\}$ it will be enough to show that for each $x \in V$ there is some triangle $\bar{A}_h$, $1 \leq h \leq 2^n$, such that the homothetic $\bar{\bar{A}}_h$ of $\bar{A}_h$ with center the upper vertex of $\bar{A}_h$ and ratio 2 contains x. If this happens we clearly have

$$\bar{M}\chi_K(x) \geq \frac{|K \cap \bar{\bar{A}}_h|}{|\bar{\bar{A}}_h|} = \frac{|\bar{A}_h|}{|\bar{\bar{A}}_h|} = \frac{1}{4} > \frac{1}{8}$$

The fact we have pointed out is easily seen by looking at the situation of the bases of the triangles $\bar{A}_1, \bar{A}_2, \bar{A}_3, \ldots, \bar{A}_{2^n}$ and their lateral sides in Figure 11. The basis $B_2 C_2$ over laps $B_1 C_1$, $B_3 C_3$ overlaps $B_2 C_2$, ... The side $M_2 B_2$ is parallel to $M_1 C_1$, the side $M_3 B_3$ is parallel to $M_2 C_2$ ... This permits us to conclude that V satisfies the above property.

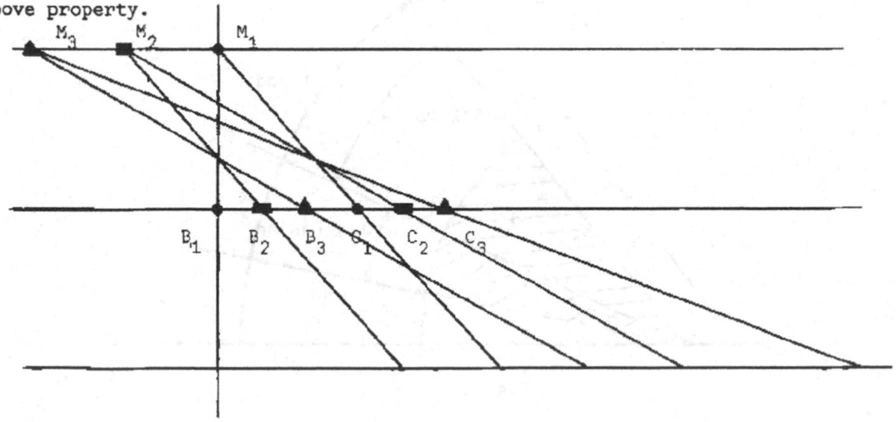

Fig. 11

REMARKS.

(1) An open problem suggested by Theorem 2.1.

Let $\{R_k\}$ be a given sequence of open rectangles. Let $\mathcal{B}$ be the B - F basis that is invariant by homothecies generated by $\{R_k\}$. Is $\mathcal{B}$ a density basis? We have produced an example, Theorem 2.1, in which $\mathcal{B}$ is not a density basis. Of course, if all rectangles R_k have the sides in the same pair of rectangular directions, then $\mathcal{B}$ is a subbasis of $\mathcal{B}_2$ and so it is a density basis. Furthermore, in this case $\mathcal{B}$ differentiates $L(1 + \log^+ L)$. The same is easily seen to hold if the number of the directions of the sides of all R_k is finite. If D_k is the length of the larger side of R_k and d_k is that of the smaller one and we have $\frac{D_k}{d_k} < M < \infty$, for all k, then $\mathcal{B}$ satisfies the Besicovitch property (cf. Remark (4) of I.1) and so $\mathcal{B}$ differentiates L^1.

One can also produce an example of a sequence of rectangles $\{R_k\}$ so that the number of side directions is infinite and $\frac{D_k}{d_k} \to \infty$ as $k \to \infty$ and still the basis $\mathcal{B}$ generated by $\{R_k\}$ is a density basis, even more $\mathcal{B}$ differentiates $L(1 + \log^+ L)$. Such a basis can be constructed as indicated in Figure 12.

Let T_1 be the open triangle determined by $I_0\ 0\ I_1$, T_2 the one determined by $I_1\ 0\ I_2$, and so on. For each T_h we define the rectangle R_h with basis $I_{h-1}\ I_h$ so that the opposite side goes through 0. We now consider the B - F basis $\mathcal{B}$ invariant by homothecies generated by $\{R_h\}_{h=1}^{\infty}$.

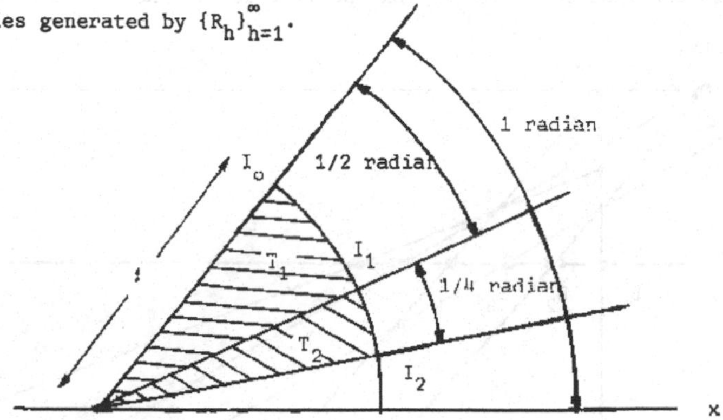

Fig. 12

In order to show that $\mathcal{B}$ differentiates $L(1 + \log^+ L)$ we proceed as follows. Observe that, if for each T_h we consider the interval P_h of Figure 13 we get

$$\frac{|T_h|}{|P_h|} = \frac{\sin \frac{1}{2^h}}{\sin \frac{1}{2^{h-1}}} \frac{\frac{1}{2}}{\cos \frac{1}{2h}} = \frac{1}{4 \cos^2 \frac{1}{2^h}} > \frac{1}{4}$$

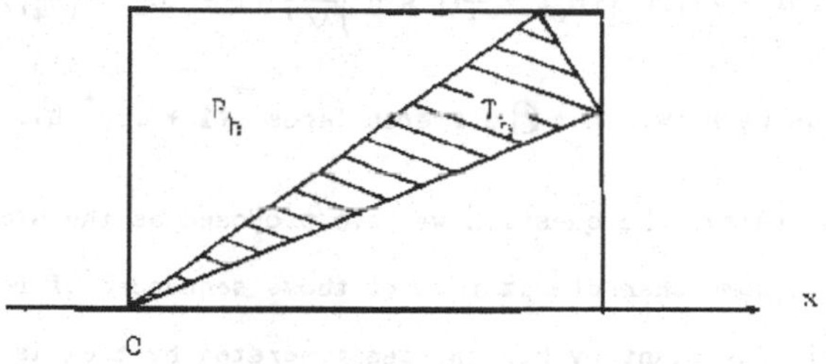

Fig. 13

So, for any $f \in L_{loc}(R^2)$ we get

$$\frac{1}{|T_h|} \int_{T_h} |f| \leq \frac{4}{|P_h|} \int_{P_h} |f|$$

In this way we obtain

$$\overline{M}f \leq 4\, M_2 f$$

where $\overline{M}$ is the maximal operator associated to the B - F basis $\bar{\mathcal{B}}$ invariant by homothecies generated by $\{T_h\}$. Now for any $B \in \mathcal{B}$ we have an element $\overline{B} \in \bar{\mathcal{B}}$ containing B

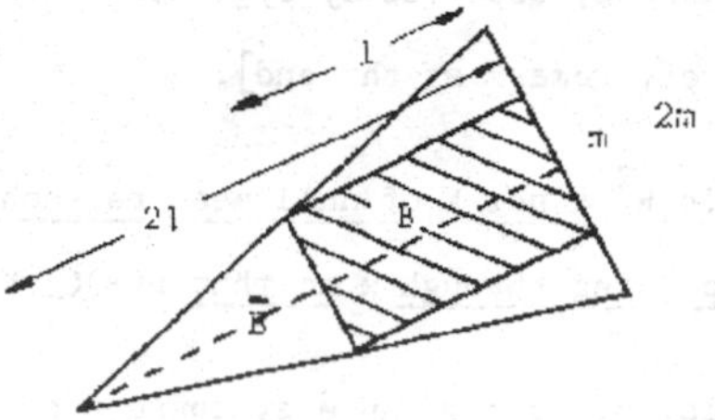

Fig. 14

and such that $|\overline{B}| = 2|B|$ and we get as before, for any $f \in L_{loc}(R^2)$

$$Mf \leqslant 2\,\overline{M}f$$

So, for any $\lambda > 0$ and any $f \in L_{loc}(R^2)$ we get

$$|\{Mf > \lambda\}| \leqslant |\{M_2 f > \frac{\lambda}{4}\}| \leqslant c \int \frac{|f|}{\lambda/4} (1 + \log^+ \frac{|f|}{\lambda/4})$$

and this implies, as we know, that $\mathcal{B}$ differentiates $L(1 + \log^+ L)$.

This result raises the question we have proposed as the starting point of this remark. Can one give some characterization of those sequences of rectangles $\{R_k\}$ so that the B - F basis invariant by homothecies generated by them is a density basis?

3. <u>THE NIKODYM SET. SOME OPEN PROBLEMS.</u>

Nikodym [1927] presented the construction of a set N, that we shall call the <u>Nikodym set</u>, contained in the unit square of R^2, such that $|N| = 1$ and for each $x \in N$ there is a straight line $r(x)$ so that $r(x) \cap N = \{x\}$. The construction of Nikodym is elementary, but extremely complicated. Such type of sets have been later on constructed, especially by Davies [1953] and Cunningham [1971,1974]. The construction we present here is based on simplifications due to these authors and utilizes the process of Theorem 1.1.

As we shall later show, this result has important consequences for the differentiation theory. This was already observed by Zygmund at the time of the original construction of Nikodym [1927, cf. remark at the end].

<u>THEOREM</u> 3.1. <u>There is in</u> R^2 <u>a set</u> K <u>of null measure such that for each</u> $x \in R^2$ <u>there is a straight line</u> $r(x)$ <u>passing through</u> x <u>so that</u> $r(x) \subset K \cup \{x\}$.

The result of Nikodym is, of course, an easy consequence of this theorem. In fact if Q is the unit square and $N = Q - K$, then $|N| = 1$ and for each $x \in N$ the line r(x) satisfies $r(x) \cap N = \{x\}$.

The proof of Theorem 3.1. is based on two lemmas.

LEMMA 3.2. *Let* R_1, R_2 *be two closed parallelograms in* R^2 *such that* $R_1 \subset R_2$. *Let* $\varepsilon > 0$ *and let* ω *be one of the two closed strips determined by* R_1. *Then there is a finite collection of closed strips* $\Omega = \{\omega_1, \omega_2, \ldots, \omega_k\}$ *such that*

(1) *For each* $i = 1, 2, \ldots, k$, $\omega_i \cap R_2 \subset \omega \cap R_2$.

(2) $R_1 \subset \bigcup_{i=1}^{k} \omega_i$.

(3) $|(\bigcup_{i=1}^{k} \omega_i) \cap (R_2 - R_1)| \leq \varepsilon$.

Proof. We shall first prove the lemma assuming that R_1 and R_2 are rectangles, A B C D and E F G H in figure 15, that R_1 has one side A D on the side E H of R_2 and that ω is the closed strip determined by the two sides of R_1 perpendicular to A D. Let $\varepsilon > 0$ be given. We draw the lines s and t as in Figure 15 perpendicular and parallel to B C respectively in such a way that the area determined by s and B I within B I J C is less than $\varepsilon/8$ and the area determined by t and B C in B I J C is also less than $\varepsilon/8$. Let u be the line parallel to t, at equal distance from t and B C, and η be the distance between t and u.

We consider the point L_1, of intersection of t and s, and the point Q_1 of intersection of u and s. Let us take a point U between P and I, at a distance ρ from P that will be conveniently fixed in a moment. For the time being it will be enough that the line UL_1, when prolongated, intersects the closed segment A D and that the angle UL_1P measures less than 45°. The line UL_1 determines a point Q_2 on u and we get a first triangle $L_1 Q_1 Q_2$. Through Q_2 we draw a line parallel to s obtaining the point L_2 on t. If the line through L_2 parallel to $L_1 Q_2$ intersects the closed segment A D we take this line, so obtaining the second triangle $L_2 Q_2 Q_3$. If this line does not intersect A D then we take instead the line L_2 D, obtaining the triangle $L_2 Q_2 Q_3$. In this way we continue. After a finite number k of steps, Q_{k+1} is for the first time within the right half of the strip ω determined by R_1.

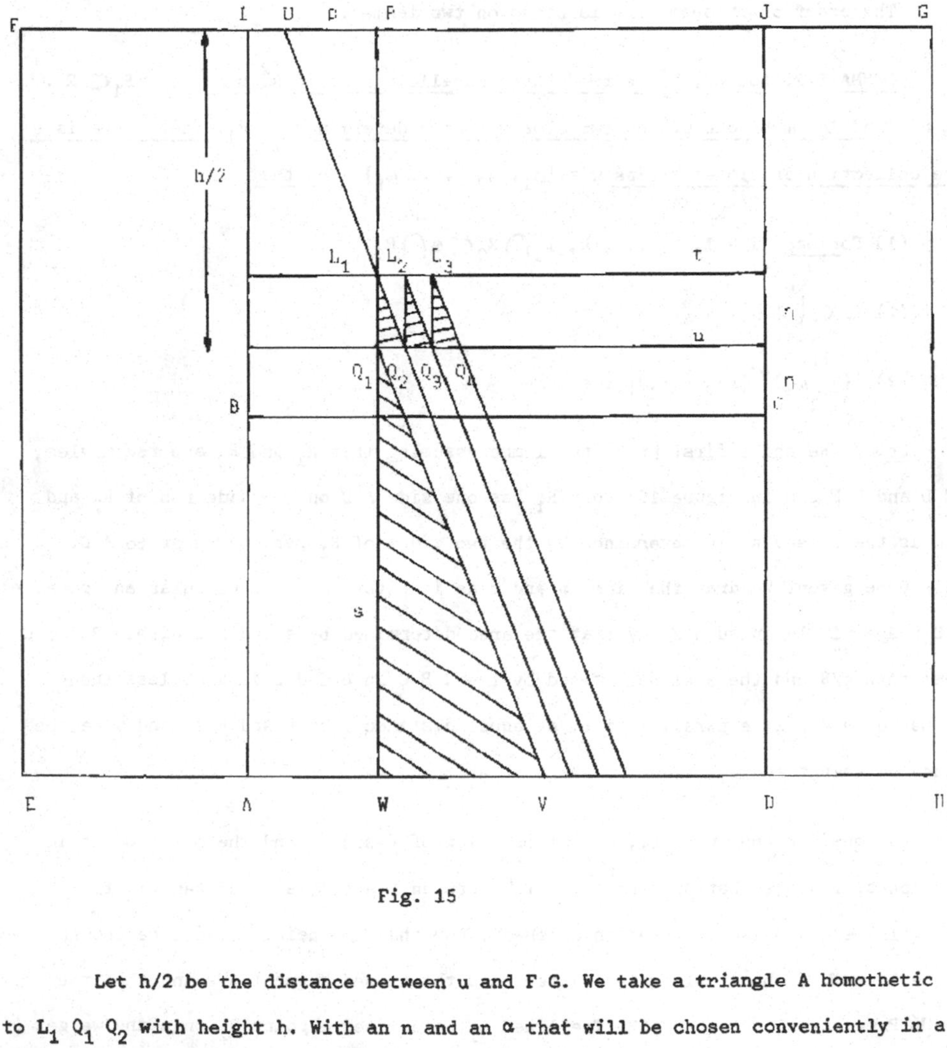

Fig. 15

Let h/2 be the distance between u and F G. We take a triangle A homothetic to $L_1\ Q_1\ Q_2$ with height h. With an n and an α that will be chosen conveniently in a moment we apply to A the process of Theorem 1.1. Our goal is to proceed in such a way that the final triangle H in that process homothetic to A (cf. page 128) be precisely $L_1\ Q_1\ Q_2$. So we have to take $\alpha^n h = \eta$. The triangles $A_1, A_2, A_3, \ldots, A_{2^n}$ of A, once they have been shifted as in Theorem 1.1. form a figure $\bar{A}_1 \cup \bar{A}_2 \cup \bar{A}_3 \cup \ldots \cup \bar{A}_{2^n}$ (the corresponding Perron tree) composed by $H = L_1\ Q_1\ Q_2$ plus additional triangles whose union has an area less than

$$2(1-\alpha)|A| = 2(1-(\tfrac{n}{h})^{1/n})\,\tfrac{1}{2}\,h^2 \tan Q_1 L_1 Q_2 \leq$$

$$\leq (1-(\tfrac{n}{h})^{1/n})\,h^2.$$

This can be made less than $\varepsilon/8k$ if n is sufficiently big. (Remember that k is the number of triangles $L_i\ Q_i\ Q_{i+1}$ we have defined). For each shifted triangle $\bar{A}_h$ we proceed to select a finite number of strips in the way indicated in Figure 16, detail of Figure 15. Let α, β be the points of intersection of the two lateral sides of $\bar{A}_h$ with F G, and γ, δ the points of intersection of these two sides with A D. We cover the closed segment $\gamma\ \delta$ with a finite number of closed segments equal in length to the segment $\alpha\ \beta$ and contained in $\gamma\ \delta$ and then we consider the closed strips obtained by joining α and β to the extreme points of these segments. By construction these strips cover the triangle Q_1 W V of Figure 15. For each one of these strips, its intersection with R_2 is contained in the strip ω determined by R_1, if only the segment $Q_1\ Q_2$ is less than UI and this is true if we choose ρ conveniently. In fact, observe that all shifted triangles $\bar{A}_1$, $\bar{A}_2$, $\bar{A}_3$, ..., $\bar{A}_{2^n}$ have their bases within $Q_1\ Q_2$ and their sides have directions within the angle $Q_1\ L_1\ Q_2$. So the left side of each one of the strips we have defined intersects F G at a distance from U to the left of U less than the length of $Q_1\ Q_2$. The area determined by the union of these strips between t and F G is less than the union of the corresponding shifted triangles, i.e. less than $\varepsilon/8k$.

We apply the same process to $L_2\ Q_2\ Q_3$, $L_3\ Q_3\ Q_4$, ..., $L_k\ Q_k\ Q_{k+1}$ and so we obtain a finite number of strips whose union covers the left half of R_1. Each one of these strips is such that its intersection with R_2 is in ω. The intersection of the union of these strips with the set between t and F G has an area less than $\varepsilon/8$. Since the area between t and B C inside ω is less than $\varepsilon/8$, the union of all the strips we have defined intersected with $R_2 - R_1$ has an area less than $\varepsilon/4$. We add to our strips the one determined by A I and W P. We take also all the strips symmetric to the ones we already have with respect to the middle line between A B and D C. So we get our collection of strips $\Omega = \{\omega_1, \omega_2, \ldots, \omega_k\}$ satisfying (1), (2) and (3).

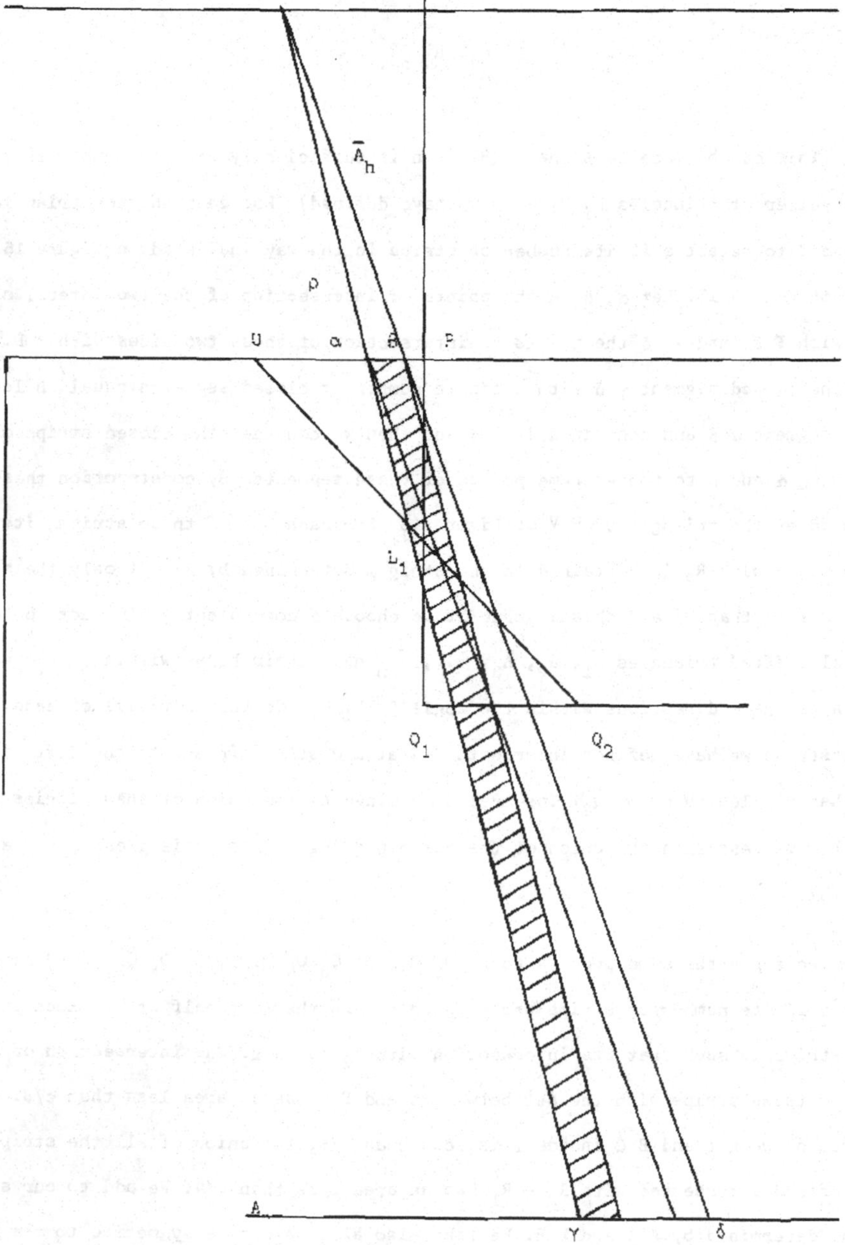

Fig. 16

We now proceed to remove the restrictions on R_1 and R_2. An affine transformation shows that the restriction imposing that R_1 and R_2 are rectangles can be easily removed. Therefore we know that the lemma holds if R_1 and R_2 are two parallelograms as in Figure 17.

Assume now that R_1 and R_2 are the parallelograms A B C D and E F G H of Figure 18.

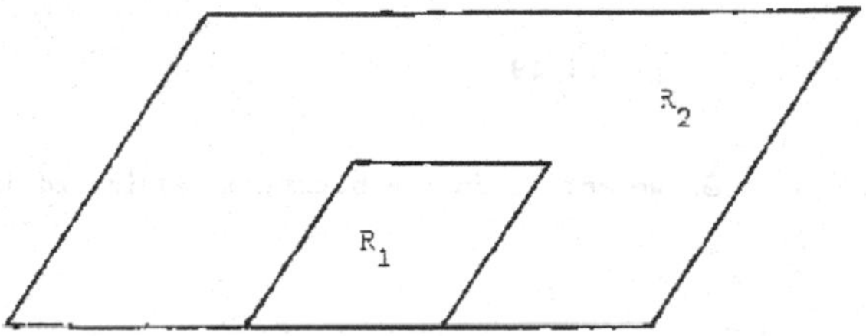

Fig. 17

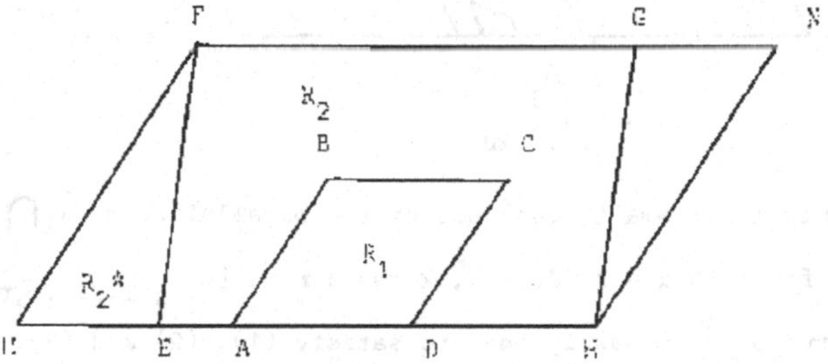

Fig. 18

We replace R_2 by $R_2^* = M\ F\ N\ H$ such that M F is parallel to A B, A D is on M H and $R_2 \subset R_2^*$. We already know that the lemma is valid for R_1 and R_2^*. It is easily seen that the same strips we obtain satisfy (1), (2) and (3) for R_1 and R_2.

Assume now that R_1 and R_2 are as in Figure 19, with A B parallel to E F and C D parallel to G H. We apply the lemma to $R_1^* = M\ B\ C\ N$ and R_2 with an $\varepsilon/2$. Each one

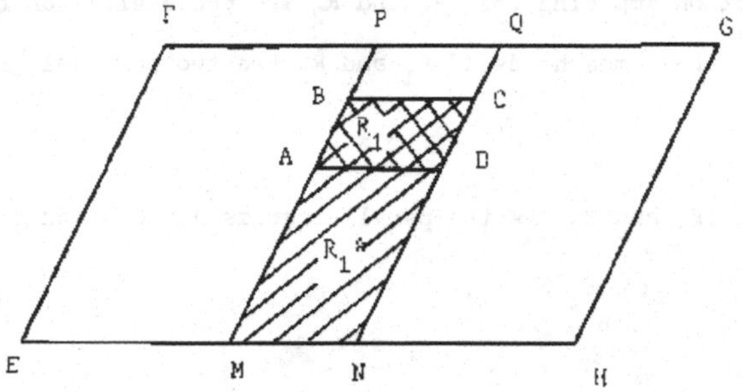

Fig. 19

of the strips $\tilde{\omega}_1, \tilde{\omega}_2, \ldots, \tilde{\omega}_k$ we get is in the situation indicated in Figure 20

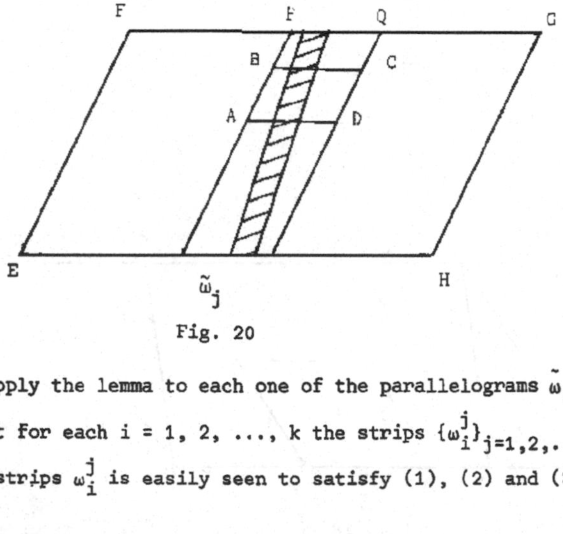

Fig. 20

So we can now apply the lemma to each one of the parallelograms $\tilde{\omega}_i \cap$ APQD and R_2 with $\varepsilon/2k$, and we get for each i = 1, 2, ..., k the strips $\{\omega_i^j\}_{j=1,2,\ldots,r_i}$. The collection Ω of all these strips ω_i^j is easily seen to satisfy (1), (2) and (3).

Finally, if R_1 and R_2 are in the general situation of the lemma one can substitute R_2 by another parallelogram R_2^*, $R_2^* \supset R_2$, with sides parallel to those of R_1 and apply the lemma to R_1 and R_2^*. The strips we obtain are also valid for R_1 and R_2.

The second lemma we are going to use is an easy consequence of the previous one.

LEMMA 3.3. Let R_1 and R_2 be two closed parallelogram such that $R_1 \subset R_2$. Let Ω be a finite collection of closed strips, $\Omega = \{\omega_1, \omega_2, \ldots, \omega_k\}$, whose union covers R_1. Let $\varepsilon > 0$ be given. Then, for each strip ω_i, $i = 1, 2, \ldots, k$ one can construct another finite collection of closed strips $\omega_i^1, \omega_i^2, \ldots, \omega_i^{j_i}$ such that, if we call $\Omega^* = \{\omega_i^j : i = 1, 2, \ldots, k, j = 1, 2, \ldots, j_i\}$, we have:

(1) $\bigcup\{\omega : \omega \in \Omega^*\} \supset R_1$.

(2) For each i and j, $\omega_i^j \cap R_2 \subset \omega_i$.

(3) $|(\bigcup\{\omega : \omega \in \Omega^*\}) \cap (R_2 - R_1)| \leq \varepsilon$.

From these two lemmas the proof of the theorem 3.1. is obtained as follows.

Proof of Theorem 3.1. For $H > 0$ let $Q(H)$ be the closed square interval centered at 0 and of side length $2H$. Let us call for brevity $Q(1) = Q$. We apply Lemma 3.2. to $R_1 = Q$ and $R_2 = Q(2)$ with an $\varepsilon_1/4 > 0$ that will be fixed later. We obtain a collection of strips Ω_1. We divide Q into four equal closed square intervals each one half the size of Q. Let us denote them by Q_1^i, $i=1,2,3,4$. Fix an i and apply Lemma 3.3. with $R_1 = Q_1^i$, $R_2 = Q(3)$, $\Omega = \Omega_1$, $\varepsilon = \varepsilon_2/4^2 > 0$. So we obtain a collection Ω^* of closed strips that we shall call Ω_2^i. Let us set

$$\Omega_2 = \bigcup_{i=1}^{4} \Omega_2^i.$$

We now divide each Q_1^i into four equal closed square intervals, each one half the size of Q_1^i. So we obtain 4^2 squares Q_2^i, $i = 1, 2, \ldots, 4^2$. Fix an i and apply again Lemma 3.3., with $R_1 = Q_2^i$, $R_2 = Q(4)$, $\Omega = \Omega_2$, $\varepsilon = \varepsilon_3/4^3$. So we obtain the collection Ω^* of the lemma that we shall denote Ω_3^i and we write

$$\Omega_3 = \bigcup_{i=1}^{4^2} \Omega_3^i.$$

And so on.

Observe that for a fixed k, the union of all strips in Ω_k^i covers the square

Q^i_{k-1}. For each $\omega \in \Omega^i_k$ we define $\hat{\omega} = \overline{\omega - Q^i_{k-1}}$ and let $K_k = \bigcup\{\hat{\omega} : \omega \in \Omega_k\}$. We have, by the construction of Ω^i_k according to Lemma 3.3.,

$$|[\bigcup\{\hat{\omega} : \omega \in \Omega^i_k\}] \cap Q(k+1)| \leq \varepsilon_k/4^k$$

and so we get $|K_k \cap Q(k)| \leq \varepsilon_k$ for each ε_k. We now define

$$K^* = \lim\inf K_k = \bigcup_{h=1}^{\infty} \bigcap_{k=h}^{\infty} K_k.$$

We choose $\varepsilon_k \to 0$ and so $|K^*| = 0$. In fact, if we fix N and h and then we take j, $j > h$, $j > N$ we obtain

$$|(\bigcap_{k=h}^{\infty} K_k) \cap Q(N)| \leq |K_j \cap Q(j)| \leq \varepsilon_j.$$

Hence

$$|[\bigcap_{k=h}^{\infty} K_k] \cap Q(N)| = 0.$$

Since this holds for each N, we get $|\bigcap_{k=h}^{\infty} K_k| = 0$ for each h and so $|K^*| = 0$.

We now show that for each $x \in Q$ there exists a straight line $r(x)$ passing through x and contained in $K^* \cup \{x\}$. Let $x \in Q$ be fixed and let $Q_n^{j(x,n)}$, $n = 1,2,3,...$ be a contracting sequence of the squares we have constructed so that $x \in Q_n^{j(x,n)}$ for each n. For $n = 1$ we take a strip ω_1 of $\Omega_1^{j(x,1)}$ containing x. For $n = 2$ there is some strip ω_2 of $\Omega_2^{j(x,2)}$ containing x and such that $\omega_2 \cap Q(2) \subset \omega_1$, and so on. For $n = k$ there is some strip ω_k of Ω_k containing x and such that $\omega_k \cap Q(k) \subset \omega_{k-1}$. So there exists a sequence of lines $\{r_k(x)\}$ passing through x and such that $r_k(x) \subset \omega_k$. Since the width of the strips ω_k tends to zero (because of the fact that $\varepsilon_k \to 0$) and one has $\omega_k \cap Q(k) \subset \omega_{k-1}$, one has that the directions of the lines $\{r_k(x)\}$ converge to the direction of a line $r(x)$ through x.

We now prove that $r(x) \subset K^* \cup \{x\}$. Let $y \in r(x)$, $y \neq x$. Then there is a natural number N such that if $n \geq N$ we have

$$y \notin Q_n^{j(x,n)} \quad \text{and} \quad y \in \mathring{Q}(n)$$

Let us take a sequence of points $\{y_k\}$ such that $y_k \in r_k(x)$, $y_k \to y$. There is an M such that, if $k \geq M$, we have

$$y_k \in Q(N) - Q_N^{j(x,N)}$$

If $i > n \geq \max(M,N)$ we can write

$$y_i \in r_i(x) \cap Q(N) \subset \omega_i \cap Q(n) \subset \omega_n$$

Since $y_i \notin Q_N^{j(x,N)}$ we also have $y_i \notin Q_n^{j(x,n)}$. Hence $y_i \in \hat{\omega}_n$. So we have proved that for a fixed $n \geq \max(M,N)$ we have $y_i \in \hat{\omega}_n$ for each $i > n$. Since $\hat{\omega}_n$ is closed, we get $y \in \hat{\omega}_n$. Hence $y \in K^*$ and this proves $r(x) \subset K^* \cup \{x\}$.

Observe now that the above process can be performed on any given square interval Q not necessarily equal to Q(1). That is, given Q there is K* such that $|K^*| = 0$ and for each $x \in Q$ there is a straight line r(x) through x so that $r(x) \subset K^* \cup \{x\}$. We apply this to $Q_1 = Q(1)$, $Q_2(2)$, ..., $Q_k(k)$, ... and we obtain K_1^*, K_2^*, ..., K_k^*, ... We now define $K = \bigcup_{k=1}^{\infty} K_k^*$ and this set satisfies the statement of the theorem.

The following result can be extracted quite easily from the preceding proof. It will be quite useful for the considerations that follow.

THEOREM 3.4. Let Q be the closed square interval centered at Q and with side length 2. The there exists a subset M of Q of full measure, i.e. $|M| = |Q|$ and a set $K^* \subset R^2$ of null measure such that for each $x \in M$ there is a straight line r(x) passing through x and contained in $K^* \cup \{x\}$ in such a way that the direction of r(x) varies in a measurable way.

<u>Proof</u>. Let us return to the proof of the theorem 3.1. The subset M of Q is going to be the complement of the union of the boundaries of all strips we have selected in that process. Clearly $|M| = |Q|$. Let us denote also by $r_k(x) \in [0,\pi)$ the angle associated to the line $r_k(x)$. We shall show that at each step k of the construction we can make a selection of lines $r_k(x)$ for $x \in M$ such that the function $x \in M \to r_k(x) \in [0,\pi)$ is a measurable function. Since we also have $r_k(x) \to r(x)$ at each $x \in M$ as $k \to \infty$ we see that r(x) is measurable on M.

Consider the strips $\omega_1^1, \omega_2^1, \omega_3^1, \ldots$ selected in the first step. To the points in $\omega_1^1 \cap M$ we assign the direction of the strip ω_1^1. To the points in $(\omega_2^1 - \omega_1^1) \cap M$ we assign the direction of the strip ω_2^1. To the points in $(\omega_3^1 - \bigcup_{j=1}^{2} \omega_j^1) \cap M$ the direction of ω_3^1. And so on. So we obtain $r_1(x)$ on M that is a step function.

Consider now $\omega_1^1 \cap M$ and the strips of the second step covering $\omega_1^1 \cap Q$. They are such that their intersections with Q are in ω_1^1. We can proceed to assign directions as above. When we now consider $(\omega_2^1 - \omega_1^1) \cap M$ and the strips of the second step covering ω_2^1 we can proceed in the same way. And so on. The second assignment $r_2(x)$ is also a step function on M. In this way we see that r(x) is measurable.

The set K* of Theorem 3.1. satisfies the statement of our theorem.

We shall now try to see what is the meaning of the results we have proved in differentiation theory. We first introduce some notation. Let r(x) be a field of directions in R^2, i.e. for each $x \in R^2$ a straight line r(x) is given. As before we denote also by r(x) the angle in $[0,2\pi)$ corresponding to this line. For each $x \in R^2$ we denote by $\mathcal{B}_r(x)$ the collection of all open rectangles of R^2 such that contain x and one of their sides is in direction r(x). We denote by $\mathcal{B}_r$ the differentiation basis $(\mathcal{B}_r(x))_{x \in R^2}$. In a paper by the author and G. Welland (Guzmán and Welland [1971]) several questions were proposed related to the differentiation properties of $\mathcal{B}_r$. With the help of Theorem 3.4. one can give the following answer to one of them. I wish to acknowledge with thanks several helpful suggestions of R.O. Davies on this subject.

<u>THEOREM</u> 3.5. <u>There exists a continuous field of directions</u> d(x) <u>in</u> R^2 <u>such</u>

that $\mathcal{B}_d$ is not a density basis.

Proof. Consider the sets M and K* and the field r(x) of Theorem 3.4. By Lusin's theorem we can take a compact subset F of M with positive measure such that r(x) restricted to F is continuous. We extend r(x) to R^2 continuously and this extension will be d(x).

We now take a sequence of open bounded sets G_k with finite measure contracting to $K^* \cap Q$. We have $|G_k| \to 0$ as $k \to \infty$. For each $x \in F$ we take r(x) = d(x) and, since $r(x) \subset K^* \cup \{x\}$, we can draw a rectangle $R_k \in \mathcal{B}_d(x)$ with d(x) as axis and with diameter less than $\frac{1}{2^k}$ such that

$$\frac{|R_k \cap G_k|}{|R_k|} > 1/3.$$

Hence, if M_k is the maximal operator associated to the basis formed by the elements of $\mathcal{B}_d$ having diameter less then $1/2^k$, we have

$$M \subset \{x \in R^2 : M_k \chi_{G_k}(x) > \tfrac{1}{3}\}$$

Hence $|\{M_k \chi_{G_k} > \frac{1}{3}\}| \geq 1$ and so, according to Theorem III. 1.1, $\mathcal{B}_d$ cannot be a density basis.

The results we have proved raise the following general problem for which one only knows several (rather trivial) partial answers.

PROBLEM. 3.6. For which fields of directions r(x) the basis $\mathcal{B}_r$ is a density basis?

REMARKS.

(1) If r is a direction field with a countable number of values, then $\mathcal{B}_r$ differentiates $L(1 + \log^+ L)$.

In fact, let $f \in L(1 + \log^+ L)$ and let

$$E_h = \{x \in R^2 : r(x) = r_h\}.$$

Let $\mathcal{B}_{r_h}$ be the basis of all rectangles with one side in direction r_h. We know that $\mathcal{B}_{r_h}$ differentiates $L(1 + \log^+ L)$. Therefore the set of points of E_h where $\mathcal{B}_r$ does not differentiate $\int f$ is a null set. Since there is a denumerable collection of sets E_h, the set where $\mathcal{B}_r$ does not differentiate $\int f$ is a null set.

(2) <u>A field that does not admit a Nikodym type of set.</u>

For each $x \in R^2$ different from 0 let $r(x)$ be the perpendicular to the line λx, $-\infty < \lambda < +\infty$. Then it is not possible to construct a set $N \subset R^2$ of positive measure such that for each $x \in N$, $x \neq 0$, one has $r(x) \cap N = \{x\}$.

We sketch the proof of this fact. Assume there exists a such set N. Let N^* be the subset of those points at which the density of N^* with respect to the basis $\mathcal{B}$ such that for each $z \in R^2$, $\mathcal{B}(z)$ is the set of all open balls centered at z. We know that $|N^*| = |N|$. Then, for some line l through the origin, the set $l \cap N^*$ has positive linear measure (i.e. if μ is the Lebesgue measure on the line l, then $\mu(l \cap N^*) > 0$). Hence there is some point $x \in l \cap N^*$, $x \neq 0$, such that the linear density of the set $l \cap N^*$ at x is 1, i.e. if I_k is any sequence of open linear intervals of l centered at x and contracting to x then one has

$$\frac{\mu(I_k \cap N^*)}{\mu(I_k)} \to 1 \quad \text{as} \quad k \to \infty$$

But for each $z \in l \cap N^*$ there is on the line $r(z)$, perpendicular to l passing through z, no other point of N^*. If k is big enough then $\mu(I_k \cap N^*) > \frac{3}{4}\mu(I_k)$ and so for the open balls $\mathcal{B}_k$ in R^2 centered at x and with diameter I_k one cannot have

$$\frac{|B_k \cap N^*|}{|B_k|} \to 1$$

as $k \to \infty$. This contradiction proves our assertion.

The fact we have shown suggests also an interesting question. There are continuous direction fields allowing a Nikodym set and some others that do not, in the sense of this remark. Can one give a criterion to distinguish them?

Another interesting question regarding this remark is as follow: For the direction field r ~~we~~ have defined at the beginning of the remark (for $x = 0$ define $r(0)$ arbitrarily), is $\mathcal{B}_r$ a density basis?

CHAPTER VI

SOME SPECIAL DIFFERENTIATION BASES

In this chapter we collect several results that are interesting from different points of view. The first section is devoted to an example of Hayes [1952] showing a basis $\mathcal{B}$ that is a density basis, and so differentiates L^∞, but has rather bad differentiation properties, since $\mathcal{B}$ does not differentiate any L^p for $1 \leq p < \infty$. Later on, in Chapter VII, we shall have occasion to see another more elaborate example, also due to Hayes [1958] showing that, given two spaces L^{p_1}, L^{p_2} with $1 \leq p_1 < p_2$, there exists a basis $\mathcal{B}$ that differentiates L^{p_2} but not L^{p_1}.

In the second section we consider several results related to B - F bases whose elements are convex. In the third section we widen a little the notion of differentiation bases in order to deal with unbounded sets and present some interesting results.

Finally in Section 4 we propose one problem.

1. AN EXAMPLE OF HAYES. A DENSITY BASIS $\mathcal{B}$ IN R^1 AND A FUNCTION g IN EACH L^p, $1 \leq p < \infty$, SUCH THAT $\mathcal{B}$ DOES NOT DIFFERENTIATE $\int g$.

The basis $\mathcal{B}$. For each $x \in [0,1)$ let $C_{n,x}$, $n = 1, 2, 3, \ldots$ the open interval centered at x and length $\frac{1}{2^n}$. Let $C^*_{n,x}$ be the open interval of center x and length $\frac{2^{n/2}}{2^n}$. Let M_n be the set of all points of $[0,1)$ of the form $\frac{p}{2^n}$ with p integer. Let P_n be the family of all open intervals centered at points of M_n and with length $\frac{1}{n2^{3n/2}}$. For each $x \in [0,1)$ let us set

$$D_{n,x} = C_{n,x} \cup [C^*_{n,x} \cap \cup\{Q : Q \in P_n\}]$$

If $x \in [0,1)$ we define $\mathcal{B}(x)$ as the sequence $\{D_{n,x}\}_{n=1}^\infty$, if $x \notin [0,1)$ then $\mathcal{B}(x)$ is the collection of all open intervals centered at x.

<u>The basis $\mathcal{B}$ is a density basis</u>. Let H_n be the number of intervals of P_n within $C^*_{n,x}$ and N_n the number of intervals of P_n with non-empty intersection with $C^*_{n,x}$. Clearly H_n and N_n are more or less $2^{n/2}$, i.e. there exist two constants c_1, $c_2 >$ > 0 independent of n such that

$$c_1\, 2^{n/2} \leq H_n \leq N_n \leq c_2\, 2^{n/2}.$$

Therefore

$$|C_{n,x}| \leq |D_{n,x}| \leq |C_{n,x}| + \frac{c_2 2^{n/2}}{n\; 2^{3n/2}} =$$

$$= |C_{n,x}|\;(1 + \frac{c_2\; 2^{n/2}\; 2^n}{n\; 2^{3n/2}}) = |C_{n,x}|\;(1+ \frac{c_2}{n}).$$

With this we easily prove that $\mathcal{B}$ is a density basis. In fact, let A be a measurable set and let $x \in A \cap [0,1)$ be such that

$$\frac{|C_{n,x} \cap A|}{|C_{n,x}|} \to 1 \quad \text{as} \quad n \to \infty$$

(Observe that almost all points of $A \cap \{[0,1)\}$ satisfy this property). Then we can write

$$1 \geq \frac{|D_{n,x} \cap A|}{|D_{n,x}|} \geq \frac{|C_{n,x} \cap A|}{|C_{n,x}|(1+c_2/n)} \to 1 \quad \text{as} \quad n \to \infty$$

Let now $x \in \{[0,1)\} - A$ be such that

$$\frac{|C_{n,x} \cap A|}{|C_{n,x}|} \to 0 \quad \text{as} \quad n \to \infty$$

(observe again that almost all points of $\{[0,1)\} - A$ satisfy this). Then subtracting from $\frac{|C_{n,x}|}{|C_{n,x}|} = 1$, we get

$$\frac{|C_{n,x} \cap A'|}{|C_{n,x}|} \to 1$$

where A' is the complement of A. Applying what we already know to A' and $x \in \{[0,1)\} \cap \cap A'$ we obtain

$$\frac{|D_{n,x} \cap A'|}{|D_{n,x}|} \to 1$$

i.e.

$$\frac{|D_{n,x} \cap A|}{|D_{n,x}|} \to 0$$

For almost all points $x \notin [0,1)$ we know that

$$\frac{|D_{n,x} \cap A|}{|D_{n,x}|} \to \chi_A(x).$$

This finishes the proof that $\mathcal{B}$ is a density basis.

<u>The function</u> g <u>in</u> L^p <u>for each</u> $p \in [1,\infty)$.

We define

$$f_n(x) = \begin{cases} n^2 & \text{if } x \in \{Q : Q \in P_n\} \\ 0 & \text{elsewhere} \end{cases}$$

$$g(x) = \sum_{n=1}^{\infty} f_n(x)$$

For each $p \in [1,\infty)$ we have

$$||g||_p \leq \sum_{n=1}^{\infty} ||f_n||_p$$

and also

$$\int f_n^p \leq \frac{2^n\, n^{2p}}{n\, 2^{3n/2}} = \frac{n^{2p-1}}{2^{n/2}}$$

Therefore $g \in L^p$.

The basis $\mathcal{B}$ does not differentiate $\int g$.

For each $x \in [0,1)$ we have for each n

$$\frac{1}{|D_{n,x}|}\int_{D_{n,x}} g \geq \frac{1}{|D_{n,x}|}\int_{D_{n,x}} f_n \geq \frac{n^2\, 2^{n/2} c_1}{n\, 2^{3n/2}} \frac{1}{\frac{1}{2^n}(1+\frac{c_2}{n})} \to \infty$$

as $n \to \infty$. Hence $\overline{D}(\int g,x) = +\infty$ at each $x \in [0,1)$.

2. BASES OF CONVEX SETS.

We shall first obtain a result related to a rather special basis of convex polygons. This theorem belongs to Guzmán and Welland [1971]. In R^2 we fix j different directions $d_1, d_2, \ldots, d_j$. Consider the system $\mathcal{B}$ of all open convex polygons such that each of their sides is parallel to one of the j fixed directions. When $j = 2$ one knows that $\mathcal{B}$ differentiates $\int f$ to f at almost every point for all $f \in L \log^+ L$, but not for $f \in L^1$. Busemann and Feller [1934] have shown that $\mathcal{B}$ is a density basis for arbitrary j. We will show that $\mathcal{B}$, for arbitrary j, differentiates $\int f$ to f at almost every point for all $f \in L \log^+ L$. The theorem depends on a geometrical lemma which is very simple in R^2. The theorem admits an extension to R^n, which is proved by the same type of geometrical arguments, although lemma 2.1. is only valid in R^n, for $n > 2$, with altered wording. For simplicity and brevity we will present the theorem only for $n = 2$.

LEMMA 2.1. *Let* $B \in \mathcal{B}$. *Among all parallelogram containing* B *there is a closed one* P(B) *of minimal area such that:*

(a) $|P(B)| \leq 2|B|$.

(b) *The sides of* P(B) *are parallel to two of the* j *given directions defining the basis*.

Proof. Let K be any open bounded convex body in R^2. An easy continuity argument shows that, among all parallelograms containing it, there is at least one of minimal area. Let $P(K)$ be any of them. We show that $|P(K)| \leq 2|K|$. To do so, it will be enough to produce a parallelogram $S \supset K$ such that $|S| \leq 2|K|$. For this, take any direction d in R^2 and consider the two supporting straight lines t_1, t_2 of K parallel to d. Take now the segment $A B$ joining any $A \in t_1 \cap \overline{K}$ and any $B \in t_2 \cap \overline{K}$. Draw the two supporting lines s_1, s_2 of K parallel to $A B$. Then it is obvious, from the convexity of K, that the parallelogram S formed by t_1, t_2, s_1, s_2 is such that $|S| \leq 2|K|$. So $|P(B)| \leq 2|B|$ is proved.

Take now any open parallelogram U of minimal area circumscribing B and assume it does not satisfy (b). Then it clearly has two opposite sides each with just one point in common with B. It is then an easy matter to show that those two sides can be simultaneously rotated around these points so to obtain a new parallelogram U^1 circumscribing B and such that $|U^1| \leq |U|$. Since U is of minimal area $|U^1| = |U|$. Rotating further we arrive at a parallelogram with two of their sides parallel to one of the j given directions. In the same way we proceed with the other two sides, obtaining a parallelogram $P(B)$ satisfying (b). This proves the lemma.

THEOREM 2.2. The differentiation basis $\mathcal{B}$ differentiates $\int f$ at almost every point and the value of the derivative is f, for all $f \in L \log^+ L$.

Proof. The theorem is a trivial corollary of the lemma and of the theorem for $j = 2$, j being the number of directions in $\mathcal{B}$. It is enough to observe that for any B, $x \in B$, we have

$$(1/|B|) \int_B |f(y)| \, dy \leq (2/|P(B)|) \int_{P(B)} |f(y)| \, dy.$$

where $P(B)$ is as in Lemma 2.1. Now the possible directions of the sides of $P(B)$ are a fixed number $O(j)$. Hence all the usual estimates from which differentiability, in the case $j = 2$, is deduced are here valid. Hence differentiability also holds here.

We shall now prove that, given any B - F basis of convex sets one can associate to it in a natural way another B - F basis $\mathcal{B}^*$ of rectangles that is in some sense, comparable to $\mathcal{B}$. This emphasizes the interest of the results we have obtained in Chapter V. The basic lemma in order to obtain this relationship is the following result due to F. John [1948]. The easy proof of it we present here is based on an idea of Córdoba and Gallego [1970].

LEMMA 2.2. *Let* K *be any open bounded convex set in* R^n. *Let* E *be any open ellipsoid of minimal volume containing* K. *Then the ellipsoid* $\frac{1}{n}$ E, *contraction by* $\frac{1}{n}$ *of* E *with respect to its center, is contained in* K.

Proof. For the sake of clarity we prove the lemma for n = 2. The existence for an open bounded convex set, of an ellipsoid of minimal area containing it is easily proved by a continuity argument.

We start with the following chain of facts that are very easy to prove. If A is an open circle in R^2 and T is a triangle inscribed in A of maximal area, then T is equilateral. This, in turn, implies that if T is an open equilateral triangle and E is an ellipse of minimal area containing it, then E is a circle. This fact easily leads to the consequence that if T is an open isosceles triangle with its height strictly less than the height of the equilateral triangle having the same basis as T, then an ellipse of minimal area containing T is obtained in the way indicated in Figure 21. First subject T to an affine transformation ϕ that leaves its basis fixed and is such that $\phi(T)$ is equilateral. Take the circunscribed circle A of $\phi(T)$, and then $\phi^{-1}(A)$ is an ellipse of minimal area containing T.

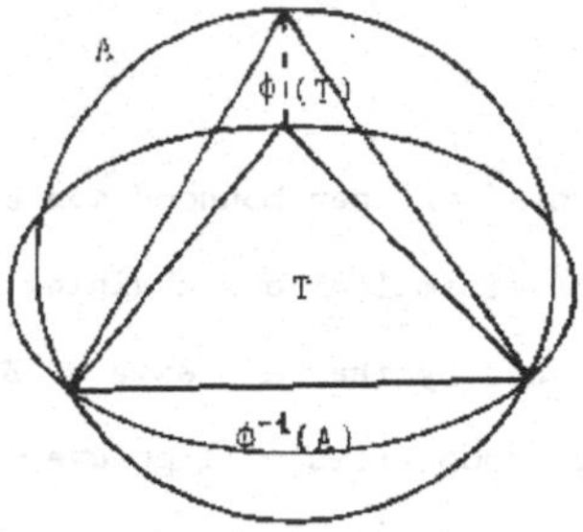

Fig. 21

One can easily deduce that $\phi^{-1}(A)$ is of area strictly smaller than the area of the circle K circumscribed to T and that $\phi^{-1}(A)$ contains the region of K above the basis of T.

With this fact we can easily prove the lemma. Assume that E is an ellipse of minimal area containing the convex set K. Assume that $\frac{1}{2}E \not\subset K$ and let $z \in (\frac{1}{2}E)^0 \cap \partial K$. Let ϕ be an affine transformation such that $\phi(E)$ is a circle. Then $\phi(z) \in (\frac{1}{2}\phi(E))^0 \cap$ $\cap \partial(\phi(K))$. Let π be one supporting line of $\phi(K)$ through $\phi(z)$. Then the situation is the one we have drawn in Figure 22.

We know that $\phi(K)$ is in the shaded region. And as we have seen, since the isosceles triangle A B C satisfies the conditions of the preceding observation, there exists

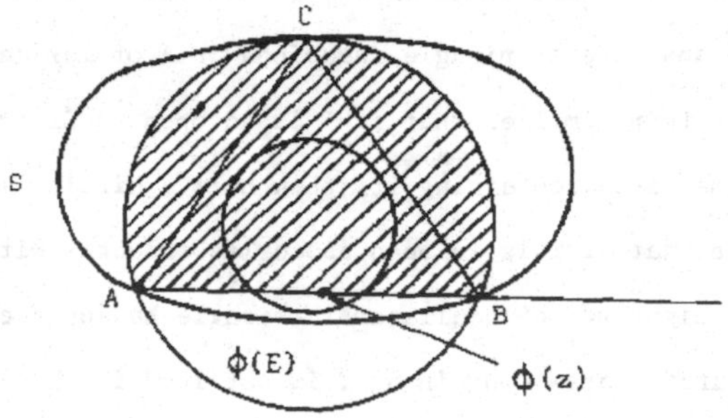

Fig. 22

an ellipse S containing the shaded region, and so $\phi(K)$. with area strictly smaller than that of $\phi(E)$. Hence $\phi^{-1}(S)$ contains K and is of smaller area than E. This contradiction proves that $\frac{1}{2}E \subset K$.

If $\mathcal{B}$ is now a B - F basis in R^n of open bounded convex sets and $B \in \mathcal{B}$, we can take each one of the ellipsoids E of minimal area containing B and then the minimal rectangular parallelepiped R determined by the main axes of E. We so obtain another B - F basis $\mathcal{B}^*$. Many differentiation properties will go over from one basis to the other. Assume for example, that $\mathcal{B}^*$ differentiates L. Then we know that for each $f \in$ $\in L$ and for almost all $x \in R^n$ we have for each sequence $\{R_k^*\} \subset \mathcal{B}^*(x)$, such that $R_k^* \to$ $\to x$,

$$\lim_{k\to\infty} \frac{1}{|R_k^*|}\int_{R_k^*} |f(y) - f(x)|\, dy = 0$$

If $\{R_k\}$ is the homologous sequence in $\mathcal{B}(x)$, then

$$\frac{1}{|R_k|}\int_{R_k} |f(y) - f(x)|dy \leq \frac{c}{|R_k^*|}\int_{R_k^*} |f(y) - f(x)|dy$$

c being a constant depending only on the dimension, since we have, with the notation used above

$$|B| \leq |R| = \pi|E| = \pi n^n|\tfrac{1}{n} E| \leq \pi n^n|B|$$

So one sees that also $\mathcal{B}$ differentiates L.

3. BASES OF UNBOUNDED SETS AND STAR-SHAPED SETS.

In this section we are going to remove the restriction imposing that the sets of a differentiation basis be bounded. Instead we shall impose finite measure. The topological contraction we have used until now will be replaced by contraction in measure. Therefore in this section a differentiation basis will be defined in the following way. For each $x \in R^2$, $\mathcal{B}(x)$ will be a collection of open sets containing x, with finite measure such that $\mathcal{B}(x)$ contains at least a sequence $\{B_k(x)\}$ of sets such that $|B_k(x)| \to 0$. We shall continue writing $B_k(x) \to x$ in order to refer to this situation. Then $\mathcal{B}=\bigcup\{\mathcal{B}(x) : x \in R^n\}$ will be a differentiation basis. For a function $f \in L^1(R^n)$ we now define correspondingly the upper and lower derivative of f at x with respect to $\mathcal{B}$ as

$$\overline{D}(\textstyle\int f,x) = \sup\{\overline{\lim_{k\to\infty}} \frac{1}{|R_k|}\int_{R_k} f : \{R_k\}\subset\mathcal{B}(x),\ R_k \to x\}$$

$$\underline{D}(\textstyle\int f,x) = \inf\{\underline{\lim_{k\to\infty}} \frac{1}{|R_k|}\int_{R_k} f : \{R_k\}\subset\mathcal{B}(x),\ R_k \to x\}$$

and we say that $\mathcal{B}$ differentiates $\int f$ when at almost every $x \in R^n$ we have

$$\overline{D}(\int f,x) = \underline{D}(\int f,x) = f(x).$$

One can state for a basis like this some interesting properties. Regarding differentiation of L^∞ one obtains the following result.

THEOREM 3.1. Let $\mathcal{B}(0)$ be the collection of all sets homothetic with center of homothecy 0 to a given open set G containing 0 such that $|G| < \infty$ and $|\partial G| = 0$, and let, for each $x \in R^n$,

$$\mathcal{B}(x) = \{x + R : R \in \mathcal{B}(0)\}$$

Then the differentiation basis $\mathcal{B} = \bigcup\{\mathcal{B}(x) : x \in R^n\}$ differentiates L^∞.

Proof. Let us call $G^h = G \cap B(0,h)$, where $B(0,h)$ is the open ball centered at 0 with radius h = 1, 2, 3, ... Let $\mathcal{B}^h$ be the basis generated starting from G^h by the same process we have used to obtain $\mathcal{B}$ starting from G. We know that $\mathcal{B}^h$ differentiates L^∞, since $\mathcal{B}^h$ is regular with respect to cubic intervals and so satisfies the Vitali property. (Moreover, it differentiates therefore L^1_{loc}). Let $f \in L^\infty$ be given. For each h there exists a null set Z_h such that for each $x \in Z_h$ and for each sequence $\{R_k^h\} \subset \mathcal{B}^h(x)$ contracting to x (topological contraction and measure contraction coincide for ^{h}) one has

$$\frac{1}{|R_k^h|}\int_{R_k^h} f \to f(x) \quad \text{as} \quad k \to \infty.$$

Let us set $Z = \bigcup_{h=1}^{\infty} Z_h$. We have $|Z| = 0$. We take $x \notin Z$ such that $|f(x)| \leq ||f||_\infty$. Then, for each fixed h, if $\{R_k^h\} \subset \mathcal{B}(x)$ and $R_k^h \to x$, we have

$$\frac{1}{|R_k^h|}\int_{R_k^h} f \to f(x) \quad \text{as} \quad k \to \infty$$

Let us take $\{R_k\} \subset \mathcal{B}(x)$, $R_k \to x$. For each fixed h let us call R_k^h the set obtained transforming G^h by the same homothecy that transforms G into R_k. Clearly $\{R_k^h\} \subset \mathcal{B}^h(x)$ and $R_k^h \to x$ as $k \to \infty$. We can write

$$\left|\frac{1}{|R_k|}\int_{R_k} f - f(x)\right| = \left|\frac{1}{|R_k|}\int_{R_k^h} f - f(x) + \frac{1}{|R_k|}\int_{R_k - R_k^h} f\right| \leq$$

$$\leq \left|\frac{|R_k^h|}{|R_k|}\frac{1}{|R_k^h|}\int_{R_k^h} f - f(x)\right| + \frac{\|f\|_\infty |R_k - R_k^h|}{|R_k|}$$

Let now $\varepsilon > 0$ be given. We choose h big enough so that

$$\frac{|G - G^h|}{|G|} = \alpha \leq \varepsilon$$

We obviously have, for each k,

$$\frac{|R_k^h|}{|R_k|} = \frac{|G^h|}{|G|} = 1 - \alpha \text{ and } \frac{|R_k - R_k^h|}{|R_k|} = \alpha \leq \varepsilon$$

Therefore, if we make $k \to \infty$ in the above inequality we obtain

$$\overline{\lim_{k\to\infty}} \left|\frac{1}{|R_k|}\int_{R_k} f - f(x)\right| \leq |(1-\alpha)f(x) - f(x)| + \varepsilon\|f\|_\infty \leq 2\varepsilon\|f\|_\infty$$

Hence, since ε can be made arbitrarily small, we get

$$\frac{1}{|R_k|}\int_{R_k} f \to f(x)$$

This concludes the proof of the theorem.

Bases of unbounded star-shaped sets can be treated by two different methods. The first one is the rotation method developped in Chapter III, by which one can obtain certain satisfactory differentation results, as we did there. The second method has been recently introduced by C.P. Calderón [1973] and consists essentially in estimating the maximal operator associated to the basis one tries to study by means of a sum of certain maximal operators, whose behavior one already knows. The problem is thus reduced to obtaining a weak type inequality for the sum of certain operators for which one knows a weak type inequality. Results in this direction appear already in the work of Cotlar [1959], E.M. Stein and N. Weiss [1969], among others.

We shall here illustrate the method with an easy typical result and refer to C.P. Calderón [1973] and Peral [1974] for more sophisticated results of the method.

Let $\{I_k\}_1^\infty$ be a sequence of open intervals in R^2 centered at the origin. Let $2D_k$ be the length of the bigger side, the horizontal one, and $2d_k$ the length of the smaller one (the vertical one). Assume $D_k \to \infty$ and $d_k \to 0$ as $k \to \infty$ is such a way that $|\bigcup_{k=1}^\infty I_k| = 1$. Let us call $G = \bigcup_{k=1}^\infty I_k$. The set G is an open star-shaped set and $|G| = 1$. We consider the basis $\mathcal{B}$ such that $\mathcal{B}(0)$ is the collection of all sets homothetic to G with center of homothetic to G with center of homothecy 0 and $\mathcal{B}(x) = \{x+R : R \in \mathcal{B}(0)\}$ for any $x \in R^2$. Then we can state the following theorem.

THEOREM 3.2. *Let $\mathcal{B}$ be the differentiation basis we have just described. Assume that*

$$\sum_{k=1}^{\infty} |I_k| \; |\log |I_k|| < \infty.$$

Then, if M is the maximal operator associated to $\mathcal{B}$, M is of weak type (1,1) and therefore $\mathcal{B}$ differentiates L^1.

Proof. Let $\mathcal{B}^k$ be the basis generated by I_k in the same way that $\mathcal{B}$ has been generated by G. Let us call M^k the maximal operator associated to $\mathcal{B}^k$. One easily gets, for example using the theorem of Besicovitch of Chapter I, that M^k is of weak type (1,1) with a constant c independent of I_k, i.e. for any $f \in L^1$ and $\lambda > 0$ we have

$$|\{M^k f > \lambda\}| \leq c \int \frac{|f|}{\lambda}$$

with c independent of f, λ and k.

We fix $x \in R^2$ and $R \in \mathcal{B}(x)$. Let $R^k \in \mathcal{B}^k(x)$ be the set obtained from I_k by the same homothecy that takes G onto R. Since $R = \bigcup_{k=1}^{\infty} R^k$ we can write, for any $f \in L^1$,

$$\frac{1}{|R|}\int_R |f| \leq \frac{1}{|R|}\sum_{k=1}^{\infty}\int_{R^k}|f| = \sum_{k=1}^{\infty}\frac{|R_k|}{|R|}\frac{1}{|R_k|}\int_{R^k}|f| =$$

$$= \sum_{k=1}^{\infty}|I_k|\frac{1}{|R_k|}\int_{R^k}|f| \leq \sum_{k=1}^{\infty}|I_k|\, M^k f(x).$$

Therefore

$$Mf(x) \leq \sum_{k=1}^{\infty}|I_k|\, M^k f(x).$$

The above quoted methods for summation of weak type inequalities lead now to the result

$$|\{Mf > \lambda\}| \leq c^* \int \left|\frac{f}{\lambda}\right|$$

where c^* is independent of f and λ. The proof given in the following lemma belongs to E.M. Stein and N.Weiss [1969].

LEMMA 3.3. Suppose that for $j = 1, 2, 3, \ldots, g_j$ is a nonnegative function on R^2 for which $|\{x : g_j(x) > s\}| \leq \frac{1}{s}$ for each $s > 0$. Let $\{c_j\}$ be a sequence of positive numbers with $\sum_j c_j = 1$ and

$$\sum_{j=1}^{\infty} c_j\, |\log c_j| = K < \infty.$$

Then, for each $s > 0$,

$$|\{x : \sum_j c_j\, g_j(x) > s\}| \leq \frac{2(K+2)}{s}$$

Proof. For each j = 1, 2, 3, ... let us define

$$v_j(x) = \begin{cases} g_j(x) & \text{if } g_j(x) < \frac{s}{2} \\ 0 & \text{if } g_j(x) \geq \frac{s}{2} \end{cases}$$

$$u_j(x) = \begin{cases} g_j(x) & \text{if } g_j(x) > \frac{s}{2c_j} \\ 0 & \text{if } g_j(x) \leq \frac{s}{2c_j} \end{cases}$$

$$m_j(x) = g_j(x) - v_j(x) - u_j(x)$$

Let $v(x) = \Sigma c_j\ v_j(x)$, $u(x) = \Sigma c_j\ u_j(x)$, $m(x) = \Sigma c_j\ m_j(x)$. Observe that, for each x, we have $v(x) < \frac{s}{2}$ and that

$$|\{x : u(x) \neq 0\}| \leq \sum_j |\{x : g_j(x) > \frac{s}{2c_j}\}| \lesssim \frac{2}{s}$$

Let us write $\lambda_j(y) = |\{x : g_j(x) > y\}|$ for $y > 0$. Observe that $y\lambda_j(y) \lesssim 1$. Then we have, since $\frac{s}{2c_j} > m_j(x) > \frac{s}{2}$ at each x,

$$\int m(x)\, dx = \int \sum_j c_j m_j(x)\, dx =$$

$$= -\sum_j c_j \int_{s/2}^{s/2c_j} y d\lambda_j(y) = \sum_j c_j \left(\int_{s/2}^{s/2c_j} \lambda_j(y)\, dy - \left[y\lambda_j(y)\right]_{s/2}^{s/2c_j} \right) \leq$$

$$\leq \sum_j c_j \int_{s/2}^{s/2c_j} \frac{1}{y} dy + 1 = \sum_j c_j\ |\log c_j| + 1 = K + 1$$

From this we get $|\{x : m(x) > \frac{s}{2}\}| \leq \frac{2(K+1)}{s}$, and finally,

$$|\{x : \Sigma c_j g_j(x) > s\}| = |\{x : m(x)+v(x)+u(x) > s\}| \lesssim$$

$$\leq |\{x : m(x) > \frac{s}{2}\}| + |\{x : u(x) \neq 0|\} + |\{x : v(x) > \frac{s}{2}\}| \leq \frac{2(K+2)}{s}$$

The application of this lemma to our situation leads immediately to the result of the theorem related to the maximal operator.

The basis $\mathcal{B}$, according to Theorem 3.1., differentiates L^∞. Therefore, using the weak type estimation we have for the maximal operator and the density of L_0^∞ (bounded functions with compact support) in L^1, we get the differentiability result of the theorem by the usual technique.

When we are given an arbitrary star-shaped open set G with $|G| = 1$ and $0 \in G$, we can try to study the maximal operator associated to the basis $\mathcal{B}$ generated by G by the same method. If G has only one infinite "peak", one can cover this peak by a sequence of intervals I_k with an economical overlapping. A condition similar to the one of Theorem 3.2. would lead us to the same result. The same is valid for a G with a finite number of peaks and even with an infinite number of peaks with an appropiate condition. We refer for details to C.P. Calderón [1973] or Peral [1974].

4. A PROBLEM.

We present here a problem suggested by Zygmund. It is related to a differentiation basis for which one does not know a sharp covering theorem, or at least one can expect to have for it a better covering theorem than the ones that are known.

Consider in R^2 the following basis $\mathcal{B}$. For each $x \in R^2$, $\mathcal{B}(x)$ will be the collection of all open intervals containing x and such that, if d is the length of the smaller side and D is the length of the larger side one has $D^2 \leq d \leq D \leq 1$.

PROBLEM. What are the differentiation properties of this basis $\mathcal{B}$?

Of course one knows that it differentiates $L(1 + \log^+ L)$, but one can expect better differentiation properties, perhaps that $\mathcal{B}$ differentiates $L(1 + \log^+ L)^{1/2}$.

This problem has been recently solved by R. Moriyón. His treatment of the problem can be seen in Appendix IV.

CHAPTER VII

DIFFERENTIATION AND COVERING PROPERTIES

In this chapter we shall try to understand the relations between the differentiation properties and the covering properties of a differentiation basis. As we have already seen (cf. especially the remarks of Section 3 in Chapter I), covering theorems have important consequences on the differentiation properties of a basis. Here we shall further develop this subject, trying to give an answer to the following question: Knowing that a basis satisfy certain differentiation properties, what can be said about its covering properties?

Although we have proceeded in Chapter III in a different way, the classical means in order to obtain differentiation theorems of measures and integrals has been the covering theorem of Vitali and certain simplifications and modifications of it. More recent are the implications in the reverse direction, from differentiation to covering properties. The first in considering such relations seems to have been R. de Possel [1936], whose main result we present in Section 1. Hayes and Pauc [1955] considerably developped the results of Possel, obtaining very interesting implications. In Section 2 we present an individual theorem that essentially originates in their work, although its proof is here simpler. Section 3 deals with a theorem due to Guzmân [1972], that extends the scope of an important theorem of Hayes and Pauc. Finally the results of this chapter allow us to present in a simple way an example due to Hayes that illustrates and motivates the problems with which we deal in this chapter.

1. THE THEOREM OF DE POSSEL

The version we present of this interesting theorem of de Possel [1936] is sufficient in order to illustrate the idea used in obtaining a covering theorem starting from a differentiation theorem. In its original version the theorem deals with a differentiation property of a class of functions, but, as we shall see, it is in fact an individual theorem. This feature permits certain useful applications.

Let $\mathcal{B}$ be a B - F basis. We shall say that a collection $\mathcal{E}$ of sets of $\mathcal{B}$ is a Vitali covering of a set A (related to $\mathcal{B}$) when for each $x \in A$ there is a sequence $\{R_k(x)\} \subset \mathcal{E}$ such that $R_k(x) \in \mathcal{B}(x)$ for each k and $\delta(R_k(x)) \to 0$ as $k \to \infty$.

THEOREM 1.1. Let $\mathcal{B}$ be a B - F basis. Assume that $\mathcal{B}$ differentiates the characteristic function of a fixed measurable set A, with $0 < |A| < \infty$. Then we have the following property:

(P) Given $\varepsilon > 0$ and given a Vitali covering $\mathcal{E}$ of A there exists $\{R_k\}_{k \geq 1} \subset \mathcal{E}$ such that

(a) $|A - \bigcup_k R_k| = 0$.

(b) $|\bigcup_k R_k - A| \leq \varepsilon$.

(c) $\sum_k |R_k| - |\bigcup_k R_k| = \int(\sum \chi_{R_k} - \chi_{\bigcup R_k}) \leq \varepsilon$.

Conversely, if $\mathcal{B}$ satisfies property (P) for each measurable set A, with $0 < |A| < \infty$, then $\mathcal{B}$ differentiates L^∞.

Proof. For the first part we proceed as follows. Let G be open, with $G \supset A$ such that $|G - A| \leq \varepsilon$ without loss of generality we can assume that all elements of $\mathcal{E}$ are contained in the set G. Otherwise we keep only those elements of $\mathcal{E}$ that satisfy this property. In this way we automatically obtain property (b). Let us take α, with $0 < \alpha < 1$, that will be chosen conveniently in a moment. We define

$$p_1 = \sup \{|R| : R \in \mathcal{E}, |A \cap R| > \alpha|R|\}$$

Since $|A| > 0$ and for each $\{R_k(x)\} \subset \mathcal{B}(x)$ with $R_k(x) \to x$ we have

$$\frac{|R_k(x) \cap A|}{|R_k(x)|} \to \chi_A(x)$$

it is clear that $p_1 > 0$. We take $R_1 \in \mathcal{B}$ such that

$$|R_1| > \frac{3}{4} p_1, \ |A \cap R_1| > \alpha |R_1|.$$

Let us call $A = A_1$ and $A_2 = A_1 - R_1$. If $|A_2| = 0$ the process of selecting R_k is finished. Otherwise we define

$$p_2 = \sup \{|R| : R \in \mathcal{B}, \ |A_2 \cap R| > \alpha |R|\}$$

and we select $R_2 \in \mathcal{B}$ such that

$$|R_2| > \frac{3}{4} p_2, \ |A_2 \cap R_2| > \alpha |R_2|$$

Define now $A_3 = A_2 - R_2$ and so on. We obtain a sequence $\{R_k\}_{k \geq 1}$ finite or infinite.

In order to see that $\{R_k\}$ satisfies (b), we first observe $(R_k \cap A_k) \cap (R_j \cap A_j) = \phi$ if $k \neq j$, and so we can write

$$\infty > |A| \geq |A \cap (\bigcup_k R_k)| = |\bigcup_k (A_k \cap R_k)| =$$

$$= \sum_k |A_k \cap R_k| \geq \alpha \sum_k |R_k|.$$

If $\{R_k\}$ is finite, we clearly have $|A - \bigcup_k R_k| = 0$. Assume that $\{R_k\}$ is an infinite sequence. Since $\sum_k |R_k| < \infty$ we have $|R_k| \to 0$ and so $p_k < \frac{4}{3} |R_k|$ is such that $p_k \to 0$. Let us call $A_\infty = A - \bigcup_k R_k$. Assume $|A_\infty| > 0$. Then, if we define

$$p_\infty = \sup \{|R| : R \in \mathcal{B}, \ |A_\infty \cap R| > \alpha |R|\}$$

we clearly have $p_\infty > 0$, $p_\infty \leq p_k$ for each k. This contradiction proves that $|A - \bigcup R_k| = 0$.

For the proof of (c) we can write, because of the inequality we have obtained

$|A| < \alpha \sum_k |R_k|$, and because of (a),

$$\sum_k |R_k| - |\bigcup_k R_k| \leq \frac{1}{\alpha} |A| - |\bigcup_k R_k| = (\frac{1}{\alpha} - 1)|A| + |A| - |\bigcup_k R_k| =$$

$$= (\frac{1}{\alpha} - 1)|A| + |A - \bigcup_k R_k| + |A \cap (\bigcup_k R_k)| - [|(\bigcup_k R_k) \cap A| +$$

$$+ |(\bigcup_k R_k) - A|] \leq (\frac{1}{\alpha} - 1)\ |A|$$

Hence, if we choose α so that $(\frac{1}{\alpha} - 1)\ |A| \leq \varepsilon$, we obtain (c).

The second part of the theorem is easy. Assume that $\mathcal{B}$ satisfies property (P) for each measurable set A with $0 < |A| < \infty$. We want to prove that $\mathcal{B}$ is a density basis, since this is equivalent, according to Theorem III 1.4, to differentiation of L^∞.

Let M be a measurable set. For a $\lambda > 0$ and $H > 0$ we define

$$A = \{x \in M' : \overline{D}(\int \chi_M, x) > \lambda > 0,\ |x| \leq H\}$$

So, for each $x \in A$ there exists $\{R_k(x)\} \subset \mathcal{B}(x)$, such that $R_k(x) \to x$ and

$$\frac{|R_k(x) \cap M|}{|R_k(x)|} > \lambda.$$

We shall prove that $|A| = 0$. If not, we take an arbitrary $\varepsilon > 0$ and apply property (P) to A with the Vitali covering

$$\mathcal{G} = (R_k(x))_{x \in A,\ k=1,2,3,\ldots}$$

and with $\varepsilon > 0$. We obtain $\{R_k\}$ satisfying (a), (b), (c). Hence, having into account that $M \subset A'$ we can write

$$|A| \leq \sum_k |R_k| \leq \frac{1}{\lambda}\sum_k |R_k \cap M| = \frac{1}{\lambda}\int_M \sum_k \chi_{R_k} =$$

$$= \frac{1}{\lambda}\int_M (\sum_k \chi_{R_k} - \chi_{\bigcup_k R_k}) + \frac{1}{\lambda}\,|(\bigcup_k R_k) \cap M| \leq \frac{\varepsilon}{\lambda} + \frac{1}{\lambda}\,|(\bigcup_k R_k) - A| \leq \frac{2\varepsilon}{\lambda}$$

Since ε is arbitrary, $|A| = 0$.

So we obtain for almost all $x \in M'$, $D(\chi_M,x) = 0$. If we apply this result to $N = M'$, we have, for almost all $x \in N' = M$, $D(\chi_N,x) = 0$. But this implies $D(\chi_M,x) = 1$. Hence $\mathcal{B}$ is a density basis.

REMARKS.

(1) A problem related to Theorem 1.1.

An interesting problem arises when one considers a B - F basis $\mathcal{B}$ that is invariant by homothecies and is a density basis. The theorem III. 1.2. afirms that for each $u \in (1,\infty)$ the function ϕ defined by

$$\phi(u) = \sup\{\frac{1}{|A|}\,|\{M\chi_A > \frac{1}{u}\}| \;:\; A \text{ bounded measurable with } |A| > 0\}$$

is finite.

The order of ϕ at infinity furnishes us with a means of distinguishing such density bases.

When one tries to go over to a covering theorem following the process of the proof of Theorem 1.1, the quantitative information contained in the possible knowledge of ϕ disappears, since that proof makes use of the density property only.

One can reasonably expect that if a basis $\mathcal{B}$ is better than another $\mathcal{B}^*$, i.e. its function ϕ is smaller than the function ϕ^* of $\mathcal{B}^*$, then its covering properties have to be better.

It would therefore be of interest to obtain a proof of de Possel's theorem that could distinguish quantitatively in the covering result corresponding to different functions ϕ. A partial answer to this question is presented in the following Sections.

(2) A covering theorem of open sets for an arbitrary basis.

Observe that if $\mathcal{B}$ is any B - F basis and O is an open set whose boundary is of null measure, then $\mathcal{B}$ differentiates χ_O. So we can apply Theorem 1.1 and obtain a covering theorem.

2. AN INDIVIDUAL COVERING THEOREM.

The following theorem, essentially due to Hayes and Pauc [1955], furnishes us with an equivalence criterion for the differentiation of an individual function $f \in L_{loc}$ in terms of a covering property. The proof we present hare is different from the original one and is published here for the first time.

THEOREM 2.1. Let $\mathcal{B}$ be a B - F basis with the density property and let $f \in L$, $f \geq 0$ be a fixed function. Then a necessary and sufficient condition in order that $\mathcal{B}$ differentiates $\int f$ is the following:

(E) Given a bounded measurable set A, given $\varepsilon > 0$ and a Vitali cover $\mathcal{V}$ of A, there exists a sequence $\{R_k\}$ such that, denoting $\chi_k = \chi_{R_k}$, $R = \bigcup_k R_k$, we have

(a) $|A - R| = 0$

(b) $|R - A| < \varepsilon$

(c) $\int(\sum\chi_k - \chi_R)f < \varepsilon$

Proof. Assume that $\mathcal{B}$ differentiates $\int f$. Let A, ε, $\mathcal{V}$, be as in property (E) of the statement of the theorem. Let $\eta > 0$ be a fixed constant that we shall choose conveniently in a moment. Let

$$A_k = \{x \in A : (1+\eta)^k \leq f(x) < (1+\eta)^{k+1}\}$$

for $k \in Z$. Since $\mathcal{B}$ differentiates $\int f$, we have, for almost every $x \in A_k$, a sequence

$$\{R_h(x)\} \subset \mathcal{B}(x), \ R_h(x) \longrightarrow x$$

such that

$$\frac{1}{|R_h(x)|} \int_{R_h(x)} f \longrightarrow f(x) \quad \text{for} \quad h \to \infty.$$

We can assume

$$\int_{R_h} f \leq (1+\eta)^{k+1} |R_h(x)| \text{ for each } h = 1,2,\ldots$$

We shall disregard the null set of A_k where the preceding is not valid and also the null set where f may be infinite. We apply de Possel's theorem to A_k with the Vitali covering obtained by means of the sets $\{R_h(x)\}$ and with an $\varepsilon_k \to 0$ that will be conveniently chosen later. We thus obtain a sequence $\{S_j^k\}_{j\geq 1}$ extracted from $(R_h(x))_{x\in A}$, $h=1,2,\ldots$, such that, if we denote $\chi_{S_j^k} \equiv \chi_j^k$ and $S^k = \bigcup_j S_j^k$, we have

(i) $|A_k - S^k| = 0$

(ii) $|S^k - A_k| < \varepsilon_k$

(iii) $\int(\sum_j \chi_j^k - \chi_{S^k}) < \varepsilon_k$.

Observe that we can also write

$$\int \sum_j \chi_j^k = \int(\sum_j \chi_j^k - \chi_{S^k}) + \int(\chi_{S^k} - \chi_{A_k}) + \int \chi_{A_k} < 2\varepsilon_k + \int \chi_{A_k}.$$

So condition (iii) can be written

$$\int \sum_j \chi_j^k \leq (1 + \gamma_k) \int \chi_{A_k}$$

where $\gamma_k > 0$ can be choosen in advance arbitrarily small.

We now prove that $\{S_j^k\}_{k,j}$ can be chosen as the family we are looking for in order to prove property (E). Observe that

$$|A - \bigcup_{k,j} S_j^k| = |\bigcup_k A_k - \bigcup_{k,j} S_j^k| \leq |\bigcup_k (A_k - \bigcup_j S_j^k)| = 0$$

and so we have (a). Also we have

$$|\bigcup_{k,j} S_j^k - A| \leq \sum_k |S^k - A_k| \leq \sum_k \varepsilon_k$$

and so we have (b) if we choose ε_k so that $\sum \varepsilon_k < \varepsilon$.

For each k we can write

$$\int f \sum_j \chi_j^k \leq (1+\eta)^{k+1} \int \sum_j \chi_j^k \leq (1+\eta)^{k+1} \quad (1+\gamma_k) \int \chi_{A_k}$$

So we can set

$$\int f \sum_{k,j} \chi_j^k = \sum_k \int f \sum_j \chi_j^k \leq \sum_k (1+\eta)^{k+1} \ (1+\gamma_k) \int \chi_{A_k} \leq$$

$$\leq \sum_k (1+\gamma_k) \ (1+\eta) \int f\chi_{A_k} \leq (1+\gamma) \ (1+\eta) \int f\chi_A$$

if only $\gamma_k \leq \gamma$ for each k. Hence

$$\int f(\sum_{k,j} \chi_j^k - \chi_{\bigcup_{k,j} S_j^k}) \leq (1+\eta)(1+\gamma) \int f(\chi_A - \chi_{\bigcup_{k,j} S_j^k}) +$$

$$+ \left[(1+\eta)(1+\gamma) - 1\right] \int f \chi_{\bigcup_{k,j} S_j^k}$$

Therefore, by choosing η, γ and ε_k conveniently we obtain (a), (b) and (c) of property (E). This concludes the first part of the theorem.

Assume now that (E) holds. Let us try to show that $\mathcal{B}$ differentiates $\int f$. We can assume that f has compact support without loss of generality. For each $r > s > 0$ we consider the set

$$A = A_{rs} = \{x \in R^n : \overline{D}(\int f,x) > r > s > f(x)\}$$

The set A is bounded and for each $x \in A$ there exists a sequence $\{R_k(x)\} \subset \mathcal{B}(x)$, $R_k(x) \to x$ such that

$$\frac{1}{|R_k(x)|} \int_{R_k(x)} f > r.$$

We apply (E) and extract from $(R_k(x))_{x \in A,\ k=1,2,\dots}$ a sequence $\{T_k\}$ satisfying (a), (b), (c). We can write, calling $T = \bigcup_k T_k$,

$$s|A| \geq \int f\chi_A = \int f(\chi_A - \sum_k \chi_k) + \int f \sum_k \chi_k \geq \int f(\chi_A - \sum_k \chi_k) +$$

$$+ r \int \sum \chi_k = \int f(\chi_A - \sum \chi_k) + r\int (\sum \chi_k - \chi_T) + r\int (\chi_T - \chi_A) + r|A|$$

(For the second inequality we have used

$$\int_{T_k} f > r \int \chi_k).$$

Given $\eta > 0$ we can choose $\varepsilon > 0$ for the application of (E) such that

$$s\,|A| \geq \eta + r\,|A|$$

i.e. $(r-s)\ |A| \leqslant \eta$. Hence $|A| = 0$. So we have proved that the set where $\overline{D}(\int f,x)$ is not $f(x)$ is of null measure. In the same way one proves that $\underline{D}(\int f,x) = f(x)$ almost everywhere. This concludes the proof of the theorem.

REMARKS.

(1) An open problem.

Theorem 2.2. gives an answer to the problem of finding a necessary and sufficient condition, in terms of covering properties, for a B - F basis $\mathcal{B}$ to differentiate L^p with a fixed $p \in (1,\infty)$. It is the following: For each bounded measurable set A, for each $\varepsilon > 0$, for each $f \in L^p$ with $f \geqslant 0$ and for each Vitali covering $\mathcal{V}$ of A, there exists a sequence $\{R_k\} \subset \mathcal{V}$ (that may depend on A, f, ε) such that, calling $\chi_{R_k} = \chi_k$,

(a) $|A - \bigcup R_k| = 0$

(b) $|\bigcup R_k - A| < \varepsilon$

(c) $\int f\ (\sum \chi_k - \chi_{\bigcup R_k}) < \varepsilon$.

Observe that if $\{R_k\}$ could be chosen so that (a), (b), (c) would be satisfy simultaneously for all functions $f \in L^p$, $f \geqslant 0$, with $||f||_p = 1$, then we would have, with $q = \frac{p}{p-1}$,

(c') $||\sum \chi_k - \chi_{\bigcup R_k}||_q < \varepsilon$.

It is an open problem whether the former necessary and sufficient condition is equivalent to the following one which drops the reference to f: For each bounded measurable set A, for each $\varepsilon > 0$, and for each Vitali cover $\mathcal{V}$ of A, there exists a sequence $\{R_k\} \subset \mathcal{V}$ such that

(a) $|A - \bigcup R_k| = 0$

(b) $|\bigcup R_k - A| < \varepsilon$

(c) $||\sum \chi_k - \chi_{\cup R_k}||_q < \varepsilon$.

In connection with this question, it is interesting to observe that the differentiation basis $\mathcal{B}$ defined in Remark (8) of I.3, that has not the Vitali property, differentiates L^1. In fact for $f \in L^1$

$$\frac{1}{|S_k(x)|} \int_{S_k(x)} f = \frac{1}{|Q_k(x)|} \int_{Q_k} f$$

where $Q_k(x)$ is the "solid" cubic interval corresponding to $S_k(x)$, i.e. $S_k(x)$ minus the cloud of rational points.

In the following section we shall prove a theorem related to the open problem we have quoted above.

3. A COVERING THEOREM FOR A CLASS OF FUNCTIONS.

In this section we present a theorem that permits us to deduce a covering property for a basis $\mathcal{B}$ from a differentiation property of $\mathcal{B}$. An easy corollary of this result will be a theorem due to Hayes and Pauc [1955] establishing a certain type of duality between covering and differentiation properties. As a matter of fact, the main result of the section, due to Guzmán [1972], is strongly inspired by their theorem and its proof.

We shall consider functions $\phi : [0,\infty) \to [0,\infty)$ satisfying : (i) $\phi(0) = 0$, (ii) ϕ is continuous, (iii) ϕ is strictly increasing, (iv) $\phi(u) \to \infty$ as $u \to \infty$. Such functions will be called <u>strength functions</u>. For such a function we consider its inverse. It is easy to see that ϕ^{-1} is also a strength function. Furthermore, if ψ and σ are strength functions, so is ψ_1 defined by $\psi_1(u) = u\sigma^{-1}(\psi(u))$ for $u \geq 0$.

We shall say that a B - F basis $\mathcal{B}$ <u>has</u> (covering) <u>strength</u> ϕ whenever for each measurable bounded set E for each Vitali cover $\mathcal{V}$ of E and for each $\varepsilon > 0$, one can select a finite collection $\mathcal{V}^* = \{S_k\}$ such that $|E - \bigcup S_k| < \varepsilon$ and

$$\int \phi\Big(\sum_k \chi_{S_k}(x) - \chi_{\bigcup_k S_k}(x)\Big)\, dx < \varepsilon.$$

(Observe that these two conditions vaguely mean that E is nearly covered by the sets S_k and that the overlap $\sum \chi_{S_k} - \chi_{\cup S_k}$ is "ϕ-small". In general, for $x \in R^n$ and for a finite sequence of measurable sets $\{A_k\}$ we shall denote $\nu(\{A_k\},x) = \sum_k \chi_{A_k}(x) - \chi_{\cup A_k}(x)$ and we call $\nu(\{A_k\},\cdot)$ the overlapping function of $\{A_k\}$).

From de Possel's theorem one can easily deduce that $\mathcal{B}$ has strength Θ, with $\Theta(u) = u$, if and only if $\mathcal{B}$ differentiates L^∞.

The following result is not difficult.

THEOREM 3.1. Let $\mathcal{B}$ be a B - F basis with strength $\phi_p(u) = u^p$ for some p, $1 < p < \infty$. Then $\mathcal{B}$ differentiates L^q, $q = \frac{p}{p-1}$.

Proof. Let $f \in L^q$. Given any $\alpha > 0$ and $H > 0$, we define

$$A = \{x \in R^n : |x| \leq H,\ |\overline{D}(\textstyle\int f,x) - f(x)| > \alpha\}$$

We shall prove that $|A| = 0$. Hence $\overline{D}(\int f,x) = f(x)$ almost everywhere. Similarly $\underline{D}(\int f,x) = f(x)$ a.e. and so the theorem will be proved.

Let $\varepsilon > 0$. We take g continuous and with compact support such that, if $h = f - g$, we have $||h||_q \leq \varepsilon$. We can then write

$$A = \{x \in R^n : |x| \leq H,\ |\overline{D}(\textstyle\int h,x) - h(x)| > \alpha\} \subset$$

$$\subset \{x \in R^n : |x| \leq H,\ \overline{D}(\textstyle\int |h|,x) > \frac{\alpha}{2}\} \cup \{|h| > \frac{\alpha}{2}\}$$

Let us call B the first set and C the second one in the last term.

We have

$$|C| = \int_C dx \leq \int \frac{|h(x)|^q}{(\frac{\alpha}{2})^q}\, dx \leq \frac{2^q \varepsilon^q}{\alpha^q}.$$

For each $x \in B$ we have a sequence $\{R_k(x)\} \subset \mathcal{B}(x)$ contracting to x such that for each k

$$\frac{1}{|R_k(x)|} \int_{R_k(x)} |h| > \frac{\alpha}{2}$$

Since $\mathcal{B}$ has strength ϕ_p, from the Vitali cover $\{R_k(x)\}_{x \in B,\ k=1,2,\ldots}$ we can extract a finite sequence $\{T_k\}$ such that $|\cup T_k - B| < \varepsilon$, $|B - \cup T_k| < \varepsilon$ and

$$\int (\sum \chi_{T_k} - \chi_{\cup T_k})^p < \varepsilon.$$

Hence we can write

$$|B| = |B \cap (\bigcup_k T_k)| + |B - \bigcup_k T_k| \leq \sum_k |T_k| + \varepsilon \leq \frac{2}{\alpha} \int |h| (\sum \chi_{T_k}) + \varepsilon =$$

$$= \frac{2}{\alpha} \int |h|(x) (\sum \chi_{T_k}(x) - \chi_{\cup T_k}(x)) dx + \frac{2}{\alpha} \int |h| \chi_{\cup T_k} + \varepsilon \leq$$

$$\leq \frac{2}{\alpha} \, ||h||_q \, ||\sum \chi_{T_k} - \chi_{\cup T_k}||_p + \frac{2}{\alpha} \, ||h||_q \, |\cup T_k|^{1/p} + \varepsilon \leq$$

$$\leq \frac{2}{\alpha} \varepsilon \, \varepsilon^{1/p} + \frac{2}{\alpha} \varepsilon (|B(0,H)| + \varepsilon)^{1/p} + \varepsilon$$

Where, in the last term $B(0,H)$ means the ball of center 0 and radius H and we have used

$$|\cup T_k| = |(\cup T_k) \cap B| + |(\cup T_k) - B| \leq |B(0,H)| + \varepsilon.$$

From this inequalities we can see that

$$|A| \leq \frac{2^p \varepsilon^p}{\alpha^p} + \frac{2}{\alpha}\varepsilon^{1+1/p} + \frac{2}{\alpha}\varepsilon(|B(0,H)| + \varepsilon)^{1/p} + \varepsilon$$

Since ε can be made arbitrarily small, $|A| = 0$. With this one concludes the proof of the theorem as we have indicated at the beginning.

We shall now prove an interesting implication in the other direction, from a differentiation property to a covering property.

For a strength function σ we shall denote by $\sigma(L_{loc})$ the set of functions $f : R^n \to R$ that are measurable and such that

$$\int_K \sigma(|f(x)|)\, dx < \infty$$

for each compact set K.

With these notions we can state the following result.

THEOREM 3.2. *Let ψ and σ be two strength functions and assume that $\psi(u) \geq u$ for $u \geq 1$. Assume also that $\mathcal{B}$ is a differentiation basis which differentiates $\int f$ for all $f \in \sigma(L_{loc})$ and has strength ψ. Then $\mathcal{B}$ has also strength ψ_1, where $\psi_1(u) =$ $= u\sigma^{-1}(\psi(u))$.*

The theorem will be proved by means of the following lemma.

LEMMA 3.3. *Let Θ be the identity function, $\Theta(u) = u$, $u \geq 0$ and ϕ a strength function such that $\phi(u)/u$ is nondecreasing. Assume $\mathcal{B}$ is a differentiation basis having strength Θ but not strength ϕ. Then there exists a set $E \subset R^n$, bounded and measurable with $|E| > 0$, two numbers $a > 0$, $b > 0$, and a Vitali cover $\mathcal{T}$ of E such that for every finite subcollection of $\mathcal{T}$, $\mathcal{T}^* = \{T_k\} \subset \mathcal{T}$ with*

$$|E - \bigcup T_k| < b, \quad \int \nu(\{T_k\},x)dx < b$$

(such $\mathfrak{F}^*$ exist, since $\mathcal{B}$ has strength θ) one has $|\bigcup\{T : T \in \mathfrak{F}^{**}\}| > b$ where $\mathfrak{F}^{**}$ is defined by

$$\mathfrak{F}^{**} = \{T \in \mathfrak{F}^* : \int_T [\nu(\mathfrak{F}^*,x)]^{-1} \phi(\nu(\mathfrak{F}^*,x))dx > a|T|\}$$

(the function inside the integral is defined to be 0 whenever $\nu(\mathfrak{F}^*,x) = 0$).

Proof of the lemma. Assume the lemma is not true. This means that for some differentiation basis $\mathcal{B}$, having strength θ but not strength ϕ one has that for every $E \subset R^n$ bounded and measurable with $|E| > 0$, for every $a > 0$, $b > 0$, and for every Vitali cover $\mathfrak{F}$ of E, it happens that for some finite sequence $\mathfrak{F}^* = \{T_k\} \subset \mathfrak{F}$, satisfying $|E - \bigcup T_k| < b$ and also $\int \nu(\{T_k\},x)\, dx < b$ one has $|\bigcup\{T : T \in \mathfrak{F}^{**}\}| \leq b$. We shall prove that under these circumstances $\mathcal{B}$ necessarily has strength ϕ, reaching so a contradiction. This will prove the lemma.

Let E, bounded and measurable with $|E| > 0$, $\varepsilon > 0$ and $\mathfrak{F}$, a Vitali cover of E be given. Take G open, $G \supset E$, $|G| < 2|E|$ and two numbers $a, b > 0$, to be fixed later. According to our assumptions, we can choose $\mathfrak{F}^* = \{T_k\}$ satisfying

$$|E - \bigcup T_k| < b, \int \nu(\{T_k\},x)dx < b, |\bigcup\{T : T \in \mathfrak{F}^{**}\}| \leq b$$

and also $\bigcup T_k \subset G$. We consider $\mathcal{S} = \mathfrak{F}^* - \mathfrak{F}^{**} = \{S_k\}$. Then, obviously,

$$|E - \bigcup S_k| \leq |E - \bigcup\{T_k : T_k \in \mathfrak{F}^*\}| + |\bigcup\{T_k : T_k \in \mathfrak{F}^{**}\}| \leq 2b.$$

Furthermore, defining again the integrand as 0 when $\nu(\mathcal{S},x) = 0$

$$\int \phi(\nu(\mathcal{S},x))\, dx = \int |\nu(\mathcal{S},x)|^{-1} \phi(\nu(\mathcal{S},x))(\sum \chi_{S_k}(x))\, dx$$

$$\leq \sum_{S_k \in \mathcal{S}} \int_{S_k} |\nu(\mathfrak{F}^*,x)|^{-1} \phi(\nu(\mathfrak{F}^*,x))dx \leq \sum_{S_k \in \mathcal{S}} a|S_k| \leq a\int \nu(\mathcal{S},x)dx +$$

$$+ a\left|\bigcup_{S_k \in \mathcal{S}} S_k\right| \leq a \int \nu(\mathcal{B}^*,x)dx + a|G| \leq ab + 2a|E|$$

having made use, for the first inequality, of the condition that $\phi(u)/u$ in non-decreasing.

If we first fix b so that $2b < \varepsilon$, and then a so that $ab + 2a\,|E| < \varepsilon$, we see that $\mathcal{B}$ has strength ϕ. So the lemma is established.

<u>Proof of the theorem</u>. Assume $\mathcal{B}$ has not strength ψ_1. Since $L^\infty_{loc}(R^n) \subset \sigma(L_{loc})$ we have, by the result of de Possel, that $\mathcal{B}$ has strength Θ. We also see that $\psi_1(u)/u$ is non-decreasing. Hence we can apply the lemma for $\phi = \phi_1$ and affirm that there is an E, bounded and measurable with $|E| > 0$, that there are a, b > 0 and a Vitali cover of E, $\mathcal{T}$ so that for every finite subcollection $\mathcal{T}^* = \{T_k\}$, with

$$|E - \cup T_k| < b, \int \nu(\mathcal{T},x)\, dx < b \text{ we have } |\{T : T \in \mathcal{T}^{**}\}| > b,$$

$\mathcal{T}^{**}$ being defined as above.

We shall try to construct a function $f \in \sigma(L_{loc})$ such that $\mathcal{B}$ does not differentiate $\int f$, reaching a contradiction. This will prove the theorem. First we take a sequence $\{b_m\}$, $b_m > 0$, $b_m \downarrow 0$, which will be fixed later conveniently. For each m = = 1, 2, ..., we take a finite sequence $\mathcal{T}^*_m = \{T^m_k\} \subset \mathcal{T}$ with $\delta(T^m_k) < \frac{1}{m}$, such that

$$|E - \cup T^m_k| < b_m, \int \psi(\nu(\mathcal{T}^*_m,x))\, dx < b_m$$

(observe that we also have $\int \nu(\mathcal{T}^*_m,x)dx < b$, since $\psi(u) \geq u$). We consider then the corresponding $\mathcal{T}^{**}_m$ and define

$$\omega_m(x) = \sigma^{-1}(\psi(\nu(\mathcal{T}^*_m,x))).$$

We now set

$$f(x) = \sup \omega_m(x).$$

So we have,

$$\int \sigma(|f(x)|)dx \leq \int \sum_m \sigma(\omega_m(x))dx \leq \sum_m \int \psi(\nu(\mathcal{B}_m^*,x))dx \leq \sum_m b_m.$$

If we take b_m sufficiently small as $m \to \infty$, then $\sum b_m < \infty$ and so $f \in \sigma(L_{loc})$. Consider now the set

$$O_m = \bigcup_m \{T_k^m \in \mathcal{B}_m^{**}\}$$

and $A = \overline{\lim_m} \sup O_m$. Then we have $|A| \geq b$. Every $x \in A$ is in some $T_k^m \in \mathcal{B}_m^{**}$ for an infinite number of m's and so, if $\mathcal{B}$ differentiates $\int f$ for $f \in \sigma(L_{loc})$ we have, since

$$\int_{T_k^m} f(x)dx > a\,|T_k^m| \text{ for } T_k^m \in \mathcal{B}_m^{**},$$

that $f(x) \geq a$ for almost all points x of A. Hence

$$\int \sigma(|f(x)|)dx \geq \sigma(a)|A| \geq \sigma(a)\, b.$$

If we choose b_m small enough as $m \to \infty$, we can make $\sum b_m < \sigma(a)b$ and this is a contradiction. So the theorem is proved.

The above mentioned theorem of Hayes and Pauc [1955] is an easy consequence of the following corollary of the theorem.

<u>COROLLARY</u> 3.4. <u>With the notation of the theorem, define</u>

$$\psi_2(u) = u\sigma^{-1}(\psi_1(u)),\ \psi_{k+1}(u) = u\sigma^{-1}(\psi_k(u)),$$

and assume $\psi_k(u) \geq u$ for all k. Then $\mathcal{B}$ has strength ψ_k for all k.

Proof. Apply the theorem k times.

THEOREM 3.5. (Hayes-Pauc). Let $\mathcal{B}$ be a B - F basis that differentiates L^p, for a fixed p with $1 < p < \infty$. Then $\mathcal{B}$ has strength ϕ_{q_1}, where $\phi_{q_1}(u) = u^{q_1}$, for each $q_1 < q = \frac{p}{p-1}$.

Proof. Since $\mathcal{B}$ differentiates L^p, it is a density basis and so has strenth θ with $\theta(u) = u$. We now apply the corollary and obtain that $\mathcal{B}$ has strength ϕ_{h_k} with $\phi_{h_k}(u) = u^{h_k}$, $h_k = \sum_0^k \frac{1}{p^k}$, for each k. Since

$$\sum_0^\infty \frac{1}{p^k} = \frac{p}{p-1} = q,$$

one gets the theorem.

REMARKS.

(1) A problem of Hayes and Pauc.

It is an open problem to find out whether $\mathcal{B}$, in the conditions of Theorem 3.5, has also strength ϕ_q. Perhaps methods of functional analysis could help in order to solver this problem. A closely related form of this problem has been recently solved by A. Córdoba [1975] (cf. Appendix I). The problem itself has been recently solved by Hayes [1975] in the affirmative

4. A PROBLEM RELATED TO THE INTERVAL BASIS.

The basis $\mathcal{B}_2$ in R^2 occupies a special place among the density bases. We already know that its maximal operator M_2 satisfies the following weak type inequality. For each measurable f in R^2 and for each $\lambda > 0$ we have

$$|\{M_2 f > \lambda\}| \leq c \int \frac{|f|}{\lambda} (1 + \log^+ \frac{|f|}{\lambda})$$

with c independent of f and λ. (Cf. Remark (3) II. 3.). Therefore, the basis $\mathcal{B}_2$ differentiates $L(1 + \log^+ L)(R^2)$ and so L^p for each p, $1 < p \leq \infty$. Therefore, according to Theorem 3.4, $\mathcal{B}_2$ has strength ϕ_q. Where $\phi_q(u) = u^q$, for each q, $1 < q < \infty$. This would also be true for any B - F basis $\mathcal{B}$ that differentiates L^p for each p with $1 < < p \leq \infty$. Since $\mathcal{B}_2$ differentiates $L(1 + \log^+ L)$ (R^2) one can ask whether $\mathcal{B}_2$ will satisfy a still better covering property. A plausible conjecture, suggested by the consideration of the Orlicz space conjugated to $L(1 + \log^+ L)$, is that $\mathcal{B}_2$ satisfies the following porperty: Let O be a measurable bounded set in R^2, $\mathcal{V} \subset \mathcal{B}_2$ a Vitali cover of O, and $\varepsilon > 0$. Then there exists a finite sequence $\{R_k\} \subset \mathcal{V}$ such that

$$|O - \bigcup R_k| < \varepsilon, \int_{\bigcup R_k} (e^{\sum \chi_{R_k}} - 1) < \varepsilon$$

It seems that $\mathcal{B}_2$ is a basis with very rich geometric properties, so that one could try to establish this conjecture by geomtric considerations.

The problem can be proposed more generally in the following way. Let Φ and Ψ be two conjugate Orlicz spaces of functions in R^n. Let $\mathcal{B}$ be a basis that differentiates Ψ. Is it true that $\mathcal{B}$ has strength ψ, where ψ is a strength function in some way associated to Ψ?

Of course, the problem of Hayes-Pauc mentioned in the remark of the previous section is a particular case of this one and should be tackled first.

The problem we state here has been successfully handled by A. Córdoba and R. Fefferman [1975] (cf. Appendix II).

5. AN EXAMPLE OF HAYES. A BASIS $\mathcal{B}$ DIFFERENTIATING L^q BUT NO L^{q_1} WITH $q_1 < q$.

In IV. 1 we have had the occasion of exhibiting a density basis that does not differentiate L^p for any $p < \infty$. It is natural to ask about the existence of a basis behaving differently with respect to different spaces L^p with $p < \infty$. The results of the previous sections allow us to construct a basis of this type. The example is due to Hayes [1952*,1958]. Let us fix a q with $1 < q < \infty$. We shall first construct a function f that belongs to L^{q_1} for each q_1 with $1 \leq q_1 < q$. Then we shall construct a basis $\mathcal{B}$ with strength $\phi_p(u) = u^p$, where $p = \frac{q}{q-1}$. So, by Theorem 3.1. $\mathcal{B}$ differentiates L^q. But we shall be able to prove that $\overline{D}(\int f,x) = +\infty$ for each x in a set of positive measure. Hence $\mathcal{B}$ does not differentiate any L^{q_1} with $q_1 < q$.

5.1. The function f.

Let $1 < q < \infty$, $p = \frac{q}{q-1}$. Let $Q = [0,1) \subset \mathbb{R}^1$ and let P_n be the set of points of $[0,1)$ of the form $\frac{k}{2^n}$, with $k = 0, 1, 2, \ldots, 2^{n-1}$. For each $x \in P_n$, let

$$I_n(x) = [x, x + \frac{1}{2^{n(p+1)+4}}).$$

Let $Q_n = \{I_n(x) : x \in P_n\}$. The 2^n intervals whose union is Q_n are disjoint, since $p > 1$. For each $I_n(x)$ let

$$I_n^*(x) = I_n(x) - \bigcup_{j=n+1}^{\infty} Q_j.$$

If N_j is the number of points of P_j, $j > n$, that are in $I_n(x)$, we have

$$N_j \leq \frac{\frac{1}{2^{n(p+1)+4}}}{1/2^j} + 1 = \frac{1}{2^{n(p+1)+4-j}} + 1$$

Hence

$$\left|\left(\bigcup_{j=n+1}^{\infty} Q_j\right) \cap I_n(x)\right| \leq \sum_{j=n+1}^{\infty} N_j \frac{1}{2^{j(p+1)+4}} < \frac{1}{2}\,|I_n(x)|$$

and therefore

$$|I_n^*(x)| > \frac{1}{2}\,|I_n(x)|.$$

Let now

$$Q_n^* = \bigcup_{x \in P_n} I_n^*(x).$$

We have

$$|Q_n^*| < |Q_n| = 2^n \frac{1}{2^{n(p+1)+4}} < \frac{1}{2^{np}}.$$

Observe that, by the definition of the sets Q_n^*, we have $Q_n^* \cap Q_m^* = \phi$ if $n \neq m$.

We now define the function f

$$f(x) = \begin{cases} n^p 2^{n(p-1)} & \text{if } x \in Q_n^* \\ 0 & \text{if } x \in R - \bigcup_1^\infty Q_n^*. \end{cases}$$

Then $f \in L^{q_1}$ if $1 \leq q_1 < q = \frac{p}{p-1}$. In fact, if $\alpha = p - q_1(p-1) > 0$ we have

$$\int f^{q_1} = \sum_{m=1}^\infty \int_{Q_m^*} f^{q_1} \leq \sum_{m=1}^\infty (m^p 2^{(p-1)m})^{q_1} \frac{1}{2^{mp}} = \sum_{m=1}^\infty m^{pq_1} \frac{1}{2^{m\alpha}} < \infty.$$

5.2. <u>The basis $\mathcal{B}$</u>. For each $x \in P_n$ we take the set $H_n^*(x) = [x, x + \frac{1}{2^n})$. Then each $y \in [0,1)$ is in exactly one of the sets $H_n^*(x)$ for a fixed n. We augment $H_n^*(x)$ with the first J_n elements $I_n^*(y)$ that exist to the right of x, if there are at least $J_n = \left[\frac{2^n}{n}\right] + 1$ such elements $I_n^*(y)$. Here $[\alpha]$ for $\alpha > 0$ means the integer part of α. Otherwise we just take the first $I_n^*(y)$ to the right of x. We call this augmented set $H_n(x)$. We clearly have

$$\delta(H_n(x)) \leq J_n \frac{1}{2^n} \leq \frac{1}{n} + \frac{1}{2^n} \to 0 \quad \text{as} \quad n \to \infty.$$

Let $n \geq 2$. We want to count how many points of P_n are in $\bigcup_{j=1}^{n-1} Q_j$, where

$$Q_j = \bigcup_{x \in P_j} I_j(x).$$

The set

$$Q_1 = \bigcup_{x \in P_1} I_1(x)$$

is the union of two disjoint intervals of length $\frac{1}{2^{p+5}}$. Each one contains at most

$$\left[\frac{\frac{1}{2^{p+5}}}{1/2^n}\right] + 1 = [2^{n-p-5}] + 1$$

points of P_n.

The set

$$Q_2 - Q_1 = \bigcup_{x \in P_2} I_2(x) - \bigcup_{x \in P_1} I_1(x)$$

is in the union of two intervals of tength $\frac{1}{2^{2(p+1)+4}}$

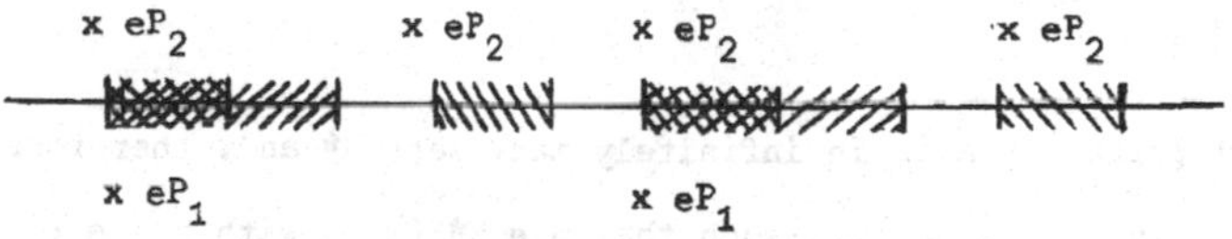

Fig. 23

each one of them containing at most $[2^{n-2(p+1)-4}]+1$ points of P_n. Similarly $Q_k - Q_{k-1}$ is in the union of 2^{k-1} intervals of length $\frac{1}{2^{k(p+1)+4}}$, each one containing at most $[2^{n-k(p+1)-4}] + 1$ points of P_n. Therefore the number of points of P_n in $\bigcup_1^{n-1} Q_j$ is not bigger than

$$2[2^{n-p-5}] + 2 + \sum_{k=2}^{n-1} (2^{k-1}[2^{n-k(p+1)-4}] + 2^{k-1}) < \frac{7}{8} 2^n.$$

Since there are 2^n points in P_n, there remain at least $\frac{1}{8} 2^n$ points of P_n that are not in $\bigcup_{j=1}^{n-1} Q_j$. This subset of P_n will be denoted by P_n^*. If a set $H_n^*(x)$ is such that $x \in P_n^*$, then we shall call it $K_n^*(x)$.

Let $x \in P_n^*$ and $1 \leq j \leq n-1$. Then the distance from x to the part of P_j that is at the right of x is at least $\frac{1}{2^n}$. Therefore we have

$$K_n^*(x) \cap \left(\bigcup_{j=1}^{n-1} Q_j \right) = \phi.$$

Let $K_n^* = \bigcup K_n^*(x)$. We have also

$$K_n^* \cap \left(\bigcup_{j=1}^{n-1} Q_j \right) = \phi.$$

We call

$$A = \limsup_{n\to\infty} K_n^* = \bigcap_{k=1}^{\infty} \bigcup_{n=k}^{\infty} K_n^*.$$

Since K_n^* contains at least $2^n/8$ disjoint intervals of length $1/2^n$, we have $|K_n^*| \geq \frac{1}{8}$ and so $|A| \geq \frac{1}{8}$

Each point $u \in A$ is in infinitely many sets K_n^* and, therefore, there exists a sequence $n_1, n_2, \ldots, n_k, \ldots$ such that $u \in K_{n_k}^*(x_{n_k})$ with $x_{n_k} \in P_{n_k}^*$. For each $u \in A$, $\mathcal{B}(u)$ will be the collection of all sets $H_{n_k}(x_{n_k})$ corresponding to these sets $K_{n_k}^*(x_{n_k}) = H_{n_k}^*(x_{n_k})$. For each $u \in R - A$, $\mathcal{B}(u)$ will be the collection of all open intervals centered at u. The collection $\mathcal{B}$ of sets in all families $\mathcal{B}(u)$ is not exactly a differentiation basis as we have defined it in Chapter II. However, the notion of upper and lower derivative is defined in the same way. With this notion we have the following.

5.3. <u>For</u> $\mathcal{B}$ <u>and</u> f <u>we have</u> $\overline{D}(\int f, u) = +\infty$ <u>for each</u> $u \in A$. Let $u \in A$. There exist , if n is big enough, $[2^n/n] + 1$ points of P_n to the right of u and so many elements $I_n(x)$, $x \in P_n$. For a such n we take $H_n(x)$ such that $u \in H_n(x)$. Then

$$|H_n(x)| \leq |H_n^*(x)| + J_n 2^{-n(p+1)-4} < 2|H_n^*(x)|.$$

Then we have, remembering that

$$Q_n^* = \bigcup_{x \in P_n} I_n^*(x), \quad |I_n^*(x)| > \frac{1}{2}|I_n(x)|, \quad \int_{H_n(x)} f \geq n^p 2^{n(p-1)}$$

$$|H_n(x) \cap Q_n^*| \geq n^p 2^{n(p-1)} \left(\left[\frac{2^n}{n}\right] + 1\right) \frac{1}{2} \frac{1}{2^{n(p+1)+4}} \geq$$

$$\geq n^{p-1} \frac{1}{2^{5+n}} \geq n^{p-1} |H_n(x)|.$$

Therefore $\overline{D}(\int f, u) = +\infty$ for each $u \in A$.

5.4. The basis $\mathcal{B}$ has strength $\phi_p(u) = u^p$.

For $\mathcal{B}$ we can define the strength, as we did in Section 3, and one easily sees that the proof of Theorem 3.5. is valid without any change. Therefore, in order to prove that $\mathcal{B}$ differentiates L^q it will be sufficient to prove that $\mathcal{B}$ has strength $\phi_p(u) = u^p$. This will be a consequence of the following three lemmas.

LEMMA A. Let $\lambda > 0$ and N a positive integer such that $\frac{1}{N} + \frac{1}{2^N} < \lambda^p$. Assume that $\mathcal{S}$ is a countable family of elements $H_n(x)$ with $n \geq N$ such that the corresponding sets $H_n^*(x)$ are disjoint. Then, for each $R \in \mathcal{S}$, we have

$$|R \cap \{x : \sum_{B \in \mathcal{S}} \chi_B(x) > 1\}|^{1/p} \leq \lambda |R|.$$

Proof. Let

$$M = \bigcup_{i=1}^{\infty} Q_i^*.$$

Since the elements $H_n(x)$ of $\mathcal{S}$ are such that the sets $H_n^*(x)$ are disjoint, we have that, if any two elements of $\mathcal{S}$ have non-empty intersection then this intersection is in the set M. So we have

$$\{x : \sum_{B\in\mathcal{S}} \chi_B(x) > 1\} \subset M$$

Hence it is enough to prove that if $R \in \mathcal{S}$, then

$$|R \cap M|^{1/p} \leq \lambda |R| .$$

Let $R = H_n(x) \in \mathcal{S}$. Since $H_n^*(x) \subset K_n^*$ and $K_n^* \cap (\bigcup_{i=1}^{n-1} Q_i^*) = \phi$, we have

$$H_n^*(x) \cap (\bigcup_{i=1}^{n-1} Q_i^*) = \phi$$

and therefore

$$H_n(x) \cap M = H_n(x) \cap (\bigcup_{i=n}^{\infty} Q_i^*).$$

We now try estimate the measure of this set. The set $H_n(x)$ contains, as we have seen before, J_n (or 1) elements $I_n^*(x)$ of Q_n^*, each one of measure less than $\frac{1}{2^{n(p+1)+4}}$. Therefore

$$|H_n(x) \cap Q_n^*| < J_n \frac{1}{2^{n(p+1)+4}} \leq \frac{1}{n2^{np}}$$

If $i > n$, then, since $Q_i^* \cap Q_n^* = \phi$, we have

$$H_n(x) \cap Q_i^* = H_n^*(x) \cap Q_i^* = H_n^* \cap Q_i .$$

Since $H_n^*(x)$ is an interval of length $1/2^n$, it contains 2^{i-n} points of P_i and so

$$|H_n(x) \cap Q_i^*| \leq |H_n^*(x) \cap Q_i| \leq 2^{i-n} \frac{1}{2^{i(p+1)+4}} < \frac{1}{2^{n+ip}}$$

Hence

$$|H_n(x) \cap (\bigcup_{i=n+1}^{\infty} Q_i^*)| < \frac{1}{2^n} \sum_{i=n+1}^{\infty} \frac{1}{2^{ip}} = \frac{1}{2^{n+np}} \sum_{i=1}^{\infty} \frac{1}{2^{ip}} <$$

$$< \frac{1}{2^{n(p+1)}}$$

and so

$$|H_n(x) \cap M| < \frac{1}{n2^{np}} + \frac{1}{2^{n(p+1)}} = \frac{1}{2^{np}} (\frac{1}{n} + \frac{1}{2^n}) \leq \lambda^p \frac{1}{2^{np}} <$$

$$< \lambda^p \; |H_n(x)|^p$$

and so the lemma is proved.

LEMMA B. The basis $\mathcal{B}$ satisfies the following property:

(R) Given $\lambda > 0$, $\varepsilon > 0$, and a Vitali cover $\mathcal{E}$ of a bounded measurable set B, there exists a sequence $\{R_k\} \subset \mathcal{E}$ such that:

(a) $|B - \bigcup R_k| = 0$

(b) $\int (\sum \chi_{R_k} - \chi_{\cup R_k}) < \varepsilon$

(c) $|\{R : R \in H(\lambda, \{R_k\})\}| < \varepsilon$, where $H(\lambda,\{R_k\})$ is the collection of elements R of $\{R_k\}$ such that

$$|R \cap \{x : \sum \chi_{R_k} > 1\}| > \lambda^p \; |R|^p.$$

Proof. Since $\mathcal{B}(x)$ for $x \notin A$ is the collection of open intervals centered at x, we can clearly assume that $B \subset A$, in order to prove the lemma.

Let

$$\lambda_n = \frac{1}{n \; 2^{n(p-1)}} + \frac{1}{2^{np}}.$$

Observe that we have

$$|H_n(x)| \leq (1 + \lambda_n)\ |H_n^*(x)|.$$

Take a positive integer N such that

$$\frac{1}{N} + \frac{1}{2^n} < \lambda^p \text{ and } \lambda_n < \varepsilon$$

for each $n > N$.

We can assume that all elements of $\mathfrak{G}$ are of the form $H_n(x)$ with $n > N$. Let $u \in B$. If $H_n(x) \in \mathcal{B}(u) \cap \mathfrak{G}$, then $H_n^*(x)$ is a half-open interval containig u. We can apply the theorem of Vitali in order to select from this sets $H_n^*(x)$ a disjoint sequence $\{H_{n_k}^*(x_{n_k})\}$ covering almostall of B. Then the corresponding sequence $\{H_{n_k}(x_{n_k})\}$ satisfies (a), (b) and (c) of property (R). In fact, (c) is a consequence of lemma A. Condition (a) is clear since

$$|B - \bigcup H_{n_k}(x_{n_m})| \leq |B - \bigcup H_{n_k}^*(x_{n_k})| = 0$$

For (b) we can set, calling $\chi_k = \chi_{H_{n_k}(x_{n_k})}$, $H = \bigcup H_{n_k}(x_{n_k})$,

$$\int (\sum \chi_k - \chi_H) = \int \sum \chi_k - |H| = \sum_k |H_{n_k}(x_{n_k})| - H \leq$$

$$\leq (1+\varepsilon) \sum_k |H_{n_k}^*(x_{n_k})| - |H| = (1+\varepsilon)\ |\bigcup H_{n_k}^*(x_{n_k})| - |H| \leq \varepsilon$$

This proves Lemma B.

<u>LEMMA C. Let $\mathcal{B}$ a basis satisfying property</u> (R) <u>of Lemma</u> B. <u>Then</u> $\mathcal{B}$ <u>has strength</u> $\phi_p(u) = u^p$.

<u>Proof</u>. Let B be a bounded measurable set, $\mathfrak{G}$ a Vitali cover of B and $\varepsilon > 0$. We choose an open set G such that $G \supset B$, $|G| < |B| + \frac{\varepsilon}{2}$. We can assume that all elements

of $\mathfrak{F}$ are contained in G.

We take $\lambda = \frac{\varepsilon^{1/p}}{|B| + \varepsilon}$ and apply property (R) obtaining $\{R_k\} \subset \mathfrak{F}$ that satisfies

$$|B - \bigcup R_k| = 0, \quad \int (\sum \chi_{R_k} - \chi_{\bigcup R_k}) < \frac{\varepsilon}{2},$$

$$|\bigcup \{R : R \in H(\lambda, \{R_k\})\}| < \frac{1}{2}\varepsilon.$$

Let $\{S_k\}$ be the elements of $\{R_k\}$ that are not in $H(\lambda, \{R_k\})$. So, for each S_k we have

$$|S_k \cap \{x : \sum_k \chi_{R_k} > 1\}| \leq \lambda^p |S_k|^p.$$

We can now write, using Minkowski's inequality:

$$\int (\sum_k \chi_{S_k} - \chi_{\bigcup S_k})^p = \int_{\sum_k \chi_{S_k} > 1} (\sum_k \chi_{S_k} - \chi_{\bigcup S_k})^p \leq \int_{\sum_k \chi_{S_k} > 1} (\sum_k \chi_{S_k})^p \leq$$

$$\leq (\sum_k (\int_{\sum_k \chi_{S_k} > 1} \chi_{S_k}^p)^{1/p})^p = (\sum_k (|S_k \cap \{\sum \chi_{S_k} > 1\}|)^{1/p})^p \leq$$

$$\leq (\sum_k \lambda |S_k|)^p = \lambda^p (\sum_k |S_k|)^p.$$

We also have

$$\sum_k |S_k| = \int \sum \chi_{S_k} \leq \int \sum \chi_{R_k} = \int (\sum \chi_{R_k} - \chi_{\bigcup R_k}) + |\bigcup R_k| \leq$$

$$\leq \frac{\varepsilon}{2} + (|B| + \frac{\varepsilon}{2}) = |B| + \varepsilon.$$

By the definition of λ we have therefore

$$\int (\sum \chi_{S_k} - \chi_{\bigcup S_k})^p \leq \varepsilon.$$

On the other hand

$$|B - \bigcup_k S_k| \leq |B - \bigcup_k R_k| + |\bigcup_k R_k - \bigcup_k S_k| \leq |B - \bigcup_k R_k| +$$

$$+ |H(\lambda, \{R_k\})| \leq \frac{\varepsilon}{2} + \frac{\varepsilon}{2} = \varepsilon.$$

With this one easily concludes the proof of the lemma, and from the three lemmas we get the fact that $\mathcal{B}$ has strength $\phi_p(u) = u^p$.

CHAPTER VIII

ON THE HALO PROBLEM

Let $\mathcal{B}$ be a B - F basis in R^n that is invariant by homothecies and satisfies the density property. According to the theorem III. 1.2. there exists a function $\phi^* : (1,\infty) \to [0,\infty)$ such that for each bounded and measurable set A and for each $u \in (1,\infty)$, one has

$$|\{M\chi_A > \frac{1}{u}\}| \leq \phi^*(u)\ |A|.$$

This suggests that we define, for any B - F basis $\mathcal{B}$ even if it is not invariant by homothecies and does not have the density property, the following function ϕ that will be called the <u>halo function of $\mathcal{B}$</u>. For each $u \in (1,\infty)$ we set

$$\phi(u) = \sup\{\frac{1}{|A|}\ |\{M\chi_A > \frac{1}{u}\}| \ : \text{A bounded, measurable, } |A| > 0\}$$

We now can say that, if $\mathcal{B}$ is invariant by homothecies then $\mathcal{B}$ is a density basis if and only if ϕ is finite at each $u \in (1,\infty)$.

If $\mathcal{B}$ is a density basis, then, for each $u > 1$ we have $|\{M\chi_A > \frac{1}{u}\}| \geq |A|$ for each A measurable with $|A| > 0$ and therefore, $\phi(u) \geq 1$. We can extend ϕ to $[0,\infty)$ by setting

$$\phi(u) = u \quad \text{for} \quad u \in [0,1].$$

In previous chapters we have seen bases whose halo functions behave rather differently. In fact, the halo function $\phi_1(u)$ of the the basis $\mathcal{B}_1$ of cubic intervals in R^n behaves like u, i.e. there exist two constants c_1 and c_2 independent of u such that

$$c_1 u \leq \phi_1(u) \leq c_2 u$$

The halo function $\phi_2(u)$ of the basis $\mathcal{B}_2$ of intervals in R^n behaves like $u(1+\log^+ u)^{n-1}$. In fact, we know from II.3, that

$$\phi_2(u) \leq c\, u(1 + \log^+ u)^{n-1}$$

The other inequality results very easily by considering $M\chi_Q$ where Q is the unit cubic interval. One easily finds

$$|\{M\chi_Q > \frac{1}{u}\}| \geq c^* u(1 + \log^+ u)^{n-1}$$

The halo function ϕ_3 of the basis $\mathcal{B}_3$ of all rectangles is infinite at each $u > 1$.

On the other hand we know that $\mathcal{B}_1$ differentiates L^1, $\mathcal{B}_2$ differentiates $L(1 + \log^+ L)^{n-1}$ (R^n) and $\mathcal{B}_3$ does not differentiates all the characteristic functions of measurable sets.

It seems clear that the order of growth of ϕ at infinity can give important information about the differentiation properties of $\mathcal{B}$. So arises the following question: Knowing the halo function ϕ of $\mathcal{B}$, invariant by homothecies, find out a minimal condition on $f \in L_{loc}(R^n)$ in order to ensure that $\mathcal{B}$ differentiates $\int f$. More precisely, the natural conjecture, looking at the picture described above, seems to be that if $\mathcal{B}$ is invariant by homothecies and ϕ is its halo function, then $\mathcal{B}$ differentiates $\phi(L)$. We shall call this the "halo conjecture". Perhaps $\mathcal{B}_1$, $\mathcal{B}_2$, $\mathcal{B}_3$ have a very particular geometric structure in order to justify the conjecture. The problem suggested by the halo function is still open.

It will be useful to look at the problem from another point of view. We know that the maximal operator M of $\mathcal{B}$ is of restricted weak type ϕ in the following sense: for each $u \in (1,\infty)$ and each A bounded measurable, with $|A| > 0$, one has

$$|\{M\chi_A > \frac{1}{u}\}| \leq \phi(u)\ |A|$$

$\phi(u)$ being the best possible constant satisfying this for all such sets A. We want to prove that M satisfies also a non-restricted weak type ϕ inequality, i.e. for each $f \in L_{loc}$ and for each $\lambda > 0$ one has

$$|\{Mf > \lambda\}| \leq c \int \phi(\frac{|f|}{\lambda})$$

In what follows we shall present some results related to the halo problem. First we deduce some easy properties of the halo function. In Section 2 we present a result of Hayes [1966], that is rather general and in Section 3 another one due to Guzmán [1973] that gives a better result for some cases. Finally we shall offer some remarks that might be useful in order to attack the problem.

1. SOME PROPERTIES OF THE HALO FUNCTION.

We consider a B - F basis that is homothecy invariant and satisfies the density property. From the definition

$$\phi(u) = \begin{cases} u & \text{if } u \in [0,1] \\ \sup \{\frac{1}{|A|}\ |\{M\chi_A > \frac{1}{u}\}| : A \text{ bounded, measurable}, |A| > 0\} \end{cases}$$

we see that ϕ is non decreasing.

When $\mathcal{B}$ is a basis of convex or star-shaped sets, one easily sees that $\phi(u) \geq u$. In fact, let $u \in (1,\infty)$. We take any set $B \in \mathcal{B}$. Let B^* be a set homothetic to B such that $B^* \supset B$ and

$$|B^*| = (\sqrt[n]{u} - \varepsilon)^n\ |B|.$$

Then $\{M\chi_B > \frac{1}{u}\} \supset B^*$ and therefore

$$\phi(u) \geqslant (\sqrt[n]{u} - \varepsilon)^n.$$

Since $\varepsilon > 0$ is arbitrary $\phi(u) \geq u$.

The following property is more interesting from the point of view of the differentiation theory.

THEOREM 1.1. Let $\mathcal{B}$ be a B - F basis that is invariant by homothecies and satisfies the density property. Let $\sigma : [0,\infty) \to [0,\infty)$ be a nondecreasing function such that $\frac{\phi(u)}{\sigma(u)} \to \infty$ for $u \to \infty$, ϕ being the halo function of $\mathcal{B}$. Then $\mathcal{B}$ does not differentiate $\sigma(L)$.

Proof. According to Remark (4) of III.3, if $\mathcal{B}$ differentiates $\sigma(L)$, then we have, for each $f \in L$ and each $\lambda > 0$

$$|\{Mf > \lambda\}| \leqslant c \int \sigma(\frac{|f|}{\lambda})$$

with c independent of f and λ.

Let us choose u_0 such that $\frac{\phi(u_0)}{\sigma(u_0)} > c$.

Then there exists a set A, measurable and bounded, with $|A| > 0$, such that

$$|\{M\chi_A > \frac{1}{{}_0}\}| > c\ \sigma(u_0)\ |A| = c \int \sigma(\frac{\chi_A}{1/u_0})$$

and this contradicts the preceding inequality taking $f = \chi_A$, $\lambda = \frac{1}{u_0}$. Therefore $\mathcal{B}$ cannot differentiate $\sigma(L)$.

2. A RESULT OF HAYES.

The following theorem constitute a good approximation to the halo conjecture.

It is essentially due to Hayes [1966] in a context a little more general and abstract than the one we set here.

THEOREM 2.1. Let $\mathcal{B}$ be a B - F basis (not necessarily invariant by homothecies). Let ϕ be the halo function of $\mathcal{B}$. Assume that ϕ is finite on $[0,\infty)$ (remember that $\phi(u) = u$ for $u \in [0,1]$). Let $\sigma : [0,\infty) \to [0,\infty)$ be a non decreasing function such that $\sigma(0) = 0$, and for some $\alpha > 1$, we have

$$\sum_{1}^{\infty} \frac{\alpha^k}{\sigma(\alpha^{k-1})} \leq 1$$

Then, for each function $f \in L$ and for each $\lambda > 0$, we have

$$|\{Mf > \lambda\}| \leq \int \phi(\sigma(\frac{4|f|}{\lambda})).$$

Proof. Assume first that $f \geq 0$. For $\lambda > 0$ let us define

$$f^{\lambda/2}(x) = \begin{cases} f(x) & \text{if } f(x) > \frac{\lambda}{2} \\ 0 & \text{if } f(x) \leq \frac{\lambda}{2}. \end{cases}$$

Then we have

$$\{Mf > \lambda\} \subset \{Mf^{\lambda/2} > \frac{\lambda}{2}\}$$

We shall now prove that, if g is a function such that its values are either 0 or bigger than 1, then we have

$$|\{Mg > 1\}| \leq \int \phi(\sigma(g)). \qquad (*)$$

With this we shall obtain

$$|\{Mf > \lambda\}| \leq |\{M(\frac{2f^{\lambda/2}}{\lambda}) > 1\}| \leq \int \phi(\sigma(\frac{2f}{\lambda}))$$

If f is not necessarily non-negative, then we can set

$$|\{Mf > \lambda\}| \leq |\{Mf^{+} > \frac{\lambda}{2}\}| + |\{Mf^{-} > \frac{\lambda}{2}\}| \leq \int \phi(\sigma(\frac{4f^{+}}{\lambda}) +$$

$$+ \int \phi(\sigma(\frac{4|f^{-}|}{\lambda})) = \int \phi(\sigma(\frac{4|f|}{\lambda})).$$

In order to prove (*), let $\alpha > 1$ be such that

$$\sum_{1}^{\infty} \frac{\alpha^{k}}{\sigma(\alpha^{k-1})} \leq 1$$

and let us call, for $k = 1, 2, \ldots,$

$$A_k = \{\alpha^{k-1} < g \leq \alpha^{k}\},\ \chi_{A_k} = \chi_k.$$

We can write

$$\{Mg > 1\} \subset \bigcup_{1}^{\infty} \{M\chi_k > \frac{1}{\sigma(\alpha^{k-1})}\}$$

since, if x is such that $M\chi_k(x) \leq \frac{1}{\sigma(\alpha^{k-1})}$ for each $k = 1, 2, \ldots,$ then, for each $B \in \mathcal{B}(x)$, we have

$$\frac{1}{|B|}\int_B g = \frac{1}{|B|} \sum_{1}^{\infty} \int_{B\cap A_k} g \leq \sum_{1}^{\infty} \frac{\alpha^{k}|B \cap A_k|}{|B|} \leq \sum_{1}^{\infty} \frac{\alpha^{k}}{\sigma(\alpha^{k-1})} \leq 1$$

Therefore

$$|\{Mg > 1\}| \leq \sum_{1}^{\infty} |\{M\chi_k > \frac{1}{\sigma(\alpha^{k-1})}\}|$$

and applying the fact that for each k we have, for $\lambda > 0$,

$$|\{M\chi_k > \lambda\}| \leq \phi(\tfrac{1}{\lambda})\ |A_k|,$$

we obtain

$$|\{Mg > 1\}| \leq \sum_{1}^{\infty} \phi(\sigma(\alpha^{k-1}))\ |A_k| \leq \int \phi(\sigma(g)).$$

This concludes the proof of the theorem.

From the theorem we have proved we easily get some interesting differentiability results. Let for example be $\phi(u) \lesssim cu$. We can take $\sigma(u) = u(1 + \log^+ u)^{1+\varepsilon}$ and we obtain

$$|\{Mf > \lambda\}| \lesssim c \int \frac{|f|}{\lambda} (1 + \log^+ \frac{|f|}{\lambda})^{1+\varepsilon}$$

This result, by routine methods (cf. Remark (3) of III. 1), shows that the corresponding basis $\mathcal{B}$ differentiates $L(1 + \log^+ L)^{1+\varepsilon}$.

Let now $\phi(u) \leq c\, u(1 + \log^+ u)$. With the same σ as before we get

$$|\{Mf > \lambda\}| \leq c \int \frac{|f|}{\lambda}\ (1 + \log^+ |\frac{f}{\lambda}|)^{2+\varepsilon}$$

and so $\mathcal{B}$ differentiates $L(1 + \log^+ L)^{2+\varepsilon}$.

As one can see, Theorem 2.1. does not give in these cases the best possible result. For $\mathcal{B}_1$ in R^2 we have $\phi(u) \leq cu$ and $\mathcal{B}_1$ differentiates L. For $\mathcal{B}_2$ in R^2, $\phi(u) \leq$ $\leq cu\,(1 + \log^+ u)$ and $\mathcal{B}_2$ differentiates $L(1 + \log^+ L)$.

In the next section we shall use another method that, for the cases indicated above, gives a finer result.

3. AN APPLICATION OF THE EXTRAPOLATION METHOD OF YANO.

In what follows we illustrate a different method for dealing with the halo problem. The results we obtain are, in a sense to be explained later, a little finer than the results derived from the theorem of Hayes. The method originates from an idea of Yano [1951] in order to handle weak type inequalities and can be seen also in Zygmund [1959,1967]. The power of the method is not exhausted in the theorem we are presenting here. It can be applied to more general situations and it can presumably given even stronger results. However we have not been able to establish with it the halo conjecture.

THEOREM 3.1. *Let $\mathcal{B}$ be a B - F differentiation basis in R^n and let ϕ be its halo function. Assume that, for some fixed $s > 0$ and for each p, with $1 < p < 2$, we have*

$$(p-1)^s \int_0^1 \lambda^{p-1}\, \phi(\tfrac{1}{\lambda})d\lambda < c < \infty$$

c being a constant independent of $p \in (1,2)$. Then $\mathcal{B}$ differentiates $L(1 + \log^+ L)^s$.

Proof. We proceed in three steps.

(a) We first prove that for each $p \in (1,2)$ we have

$$||M\chi_K||_p \leq \frac{c}{(p-1)^s}\, ||\chi_K||_p$$

where K is any bounded measurable set, and c is a constant, not necessarily the same as the previous c, independent of K and p. In fact

$$||M\chi_K||_p^p = \int_0^1 p\, \lambda^{p-1}\, |\{M\chi_K > \lambda\}|d\lambda \leq |K|\int_0^1 p\lambda^{p-1}\phi(\tfrac{1}{\lambda})d\lambda$$

and, by the hypothesis about ϕ, we get

$$||M\chi_K||_p^p \leq \frac{c}{(p-1)^s}\, ||\chi_K||_p^p$$

and from this inequality we easily get, with another constant

$$||M\chi_K||_p \leq \frac{c}{(p-1)^s} ||\chi_K||_p$$

(b) We now prove the following: Let X be any bounded measurable set. If $f \in L(1 + \log^+ L)^s$, then

$$\int_X Mf \leq c(1 + |X|) + c \int_X |f| \ (1 + \log^+|f|)^s$$

c being a constant independent of X and f. In order to show this inequality we can assume $f \geq 0$, since $Mf = M|f|$. Let us call

$$E_0 = \{x \in X : 0 \leq f(x) < 1\}$$

$$E_k = \{x \in X : 2^{k-1} \leq f(x) < 2^k\} \text{ for } k = 1, 2, 3, \ldots.$$

$$f_k = f|_{E_k}, \ |E_k| = e_k, \ k = 0, 1, 2, 3, \ldots$$

$$p_k = 1 + \frac{1}{k+1}, \ k = 1, 2, 3, \ldots, \ q_k = \frac{p_k}{p_k - 1}.$$

We can write

$$\int_X Mf < \sum_{k=0}^{\infty} \int_X Mf_k = \int_X Mf_0 + \sum_{k=1}^{\infty} \int_X Mf_k \leq |X| +$$

$$+ \sum_{k=1}^{\infty} ||Mf_k||_{p_k} ||\chi_X||_{q_k} \leq |X| + (1+|X|) \sum_{k=1}^{\infty} ||Mf_k||_{p_k} \leq$$

$$\leq |X| + (1+|X|) \sum_{k=1}^{\infty} 2^k ||M\chi_{E_k}||_{p_k} \leq |X| +$$

$$+ c(1+|X|) \sum_{k=1}^{\infty} 2^k \frac{1}{(p_k - 1)^s} e_k^{1/p_k} =$$

$$= |X| + c(1 + |X|) \sum_{k=1}^{\infty} 2^k (k+1)^s e_k^{(k+1)/(k+2)}$$

The sum of the terms in the series above corresponding those e_k for which $e_k \leq 3^{-k}$ is finite, since

$$\sum_{k=1}^{\infty} 2^k (k+1)^s \, 3^{-k\frac{k+1}{k+2}} < \infty.$$

If $e_k > 3^{-k}$, then we have

$$e_k^{-\frac{1}{k+2}} < 3^{\frac{k}{k+2}} < 3$$

and so

$$2^k (k+1)^s e_k^{\frac{k+1}{k+2}} < 2^k (k+1)^s e_k 3$$

Hence

$$\int_X Mf \leq c(1+|X|) + c\int_X |f| \, (1 + \log^+|f|)^s.$$

(c) The preceding inequality easily leads to the differentiation statement of the theorem. Let $f \in L(1 + \log^+ L)^s$, $f \geq 0$ and $\lambda > 0$. We try to estimate $\{x \in X : Mf(x) > \lambda\}$, but we first apply the inequality we have obtained to ρf with $\rho > 0$. After dividing by ρ we get

$$\int_X Mf \leq \frac{c}{\rho} (1 + |X|) + c\int_X |f| (1 + \log^+(\rho f))^s$$

So we obtain

$$|\{x \in X : Mf(x) > \lambda\}| \leq \int_X \frac{Mf}{\lambda} \leq \frac{c}{\rho\lambda}(1+|X|) + \frac{c}{\lambda}\int_X |f|\ (1+\log^+(\rho f))^s \qquad (*)$$

This permits us to apply Theorem III.1.1. to prove that $\mathcal{B}$ is a density basis. We now take a non decreasing sequence of simple functions $\{f_k\}$ such that $f_k \to f$ pointwise. Let us call $g_k = f - f_k$. Observe that, for fixed ρ and λ,

$$\frac{c}{\lambda}\int_X g_k(1 + \log^+(\rho g_k))^s \to 0.$$

We can write

$$A = \{x \in X : |\overline{D}(\textstyle\int f,x) - f(x)| > \lambda\} = \{x \in X : |\overline{D}(\textstyle\int g,x) -$$

$$- g_k(x)| > \lambda\} \subset \{x \in X : \overline{D}(\textstyle\int g_k,x) > \frac{\lambda}{2}\} \subset$$

$$\subset \{x \in X : g_k(x) > \frac{\lambda}{2}\}$$

The above inequality (*) easily leads to $|A| = 0$. In the same way

$$|B| = |\{x \in X : |\underline{D}(\textstyle\int f,x) - f(x)| > \lambda\}| = 0$$

and this proves that $\mathcal{B}$ differentiates $L(1 + \log^+ L)^s$.

4. SOME REMARKS ON THE HALO PROBLEM.

The following remarks are perhaps of interest for the solution of the halo problem, since they suggest some possible ways of handling it.

THEOREM 4.1. (a) If there exists a density B - F basis that is homothecy invariant and such that for its halo function ϕ we have

$$\frac{\phi(2u)}{\phi(u)} \to \infty \quad \text{for} \quad u \to \infty$$

then the halo conjecture is false.

(b) If the halo conjecture is true, then each density B - F basis $\mathcal{B}$, that is homothecy invariant, is such that $\mathcal{B}$ differentiates L^p for some $p < \infty$.

Proof. (a) If the halo conjecture were true, $\mathcal{B}$ would differentiate $\phi(L)$ and therefore also $\sigma(L) = \phi(\frac{L}{2})$, since $\mathcal{B}$ differentiates $\int f$ if and only if $\mathcal{B}$ differentiates $\int 2f$.

But we have

$$\frac{\phi(u)}{\phi(\frac{u}{2})} = \frac{\phi(u)}{\sigma(u)} \to \infty$$

for $u \to \infty$. Therefore, by what we have seen in Section 1, $\mathcal{B}$ does not differentiate $\sigma(L)$. This contradiction proves (a).

(b) According to (a), if the halo conjecture is true, then for $\mathcal{B}$ one has

$$\phi(2u) \leqslant c\, \phi(u)$$

where c is a constant independent of u. Therefore, since $\phi(u)$ is non decreasing, we have, if k is an integer bigger than 1,

$$\phi(2^k) \leqslant c\phi(2^{k-1}) \leqslant c^2\phi(2^{k-2}) \leqslant \dots \leqslant c^k\phi(1) = c^k$$

Hence, if $c = 2^p$, we get $\phi(2^k) \leqslant 2^{pk}$, and so, if $2^{k-1} \leqslant u < 2^k$, we obtain

$$\phi(u) \leqslant \phi(2^k) \leqslant 2^{kp} \leqslant 2^p u^p$$

According to Theorem 2.1., applied with $\sigma(u) = u \log^{1+\varepsilon}(1+u)$ we get that $\mathcal{B}$ differentiates at least $L^{p+\eta}$ for each $\eta > 0$.

Therefore, in order to disprove the halo conjecture, it would be sufficient to exhibit a density B - F basis, that is invariant by homothecies and does not differentiate any L^p with $p < \infty$. In Chapter VI we have constructed a density basis that does not differentiate any L^p with $p < \infty$, but this basis is not of the type required here.

For a counterexample to the halo conjecture one could try to construct a B - F basis invariant by homothecies and such that its halo function behaves at infinity like e^u.

The halo conjecture in the case $\psi(u) \sim u$ at infinity has been recently proved to be true by R. Moriyón [1975] (cf. Appendix III). Some other results in this direction, although not so sharp can be seen in the Appendix II by A. Córdoba and R. Fefferman. The theorem proved by Moriyón contains also information related to several questions treated in the previous chapters of these notes.

APPENDIX I

On the Vitali covering properties of a differentiation basis

by Antonio Córdoba

This appendix is related to the questions treated in Chapter VII. It is shown in Theorem 1 that certain differentiation properties of a basis $\mathcal{B}$ can be characterized by means of a covering property of the type proposed in VII.3. The technique introduced here allows us also to give some partial results about the halo problem of Chapter VIII.

Definition. Let $\mathcal{B}$ a B-F basis in $\mathbb{R}^n$. We shall say that $\mathcal{B}$ has the covering property V_q if there exists a constant C such that for every measurable bounded set E, every $\mathcal{B}$-Vitali covering V of E and any $\varepsilon > 0$, one can select a sequence $\{R_k\}$ V with the properties:

i) $|E - \bigcup R_k| = 0$, $|\bigcup R_k - E| \leq \varepsilon$

ii) $||\Sigma \chi_{R_k}||_q \leq C|E|^{1/q}$.

The purpose of this paper is to relate the following two properties of a differentiation basis:

(1) $\mathcal{B}$ differentiates $\int f$ for all $f \in L^p_{loc}(\mathbb{R}^n)$

(2) $\mathcal{B}$ has the covering property V_q, $1/p + 1/q = 1$.

For the particular case $q = 1$, $p = \infty$ the equivalence of (1) and (2) is due to de Possel [1936]. The implication (2) ⟹ (1) is well-known, and Hayes and Pauc [1955] proved that if $\mathcal{B}$ differentiates $\int f$ for all $f \in L^p_{loc}(\mathbb{R}^n)$, then $\mathcal{B}$ has the covering property V_{q_1} for all $q_1 < q$. In Theorem 1, we prove that for a basis $\mathcal{B}$ invariant by translations, the properties (1) and (2) are equivalent.

Theorem 1. Let $\mathcal{B}$ be a B-F basis that is invariatn by translations. Then the

<u>two following statements are equivalent:</u>

(1) $\mathcal{B}$ <u>differentiates</u> $\int f$ <u>for all</u> $f \in L^p_{loc}(\mathbb{R}^n)$

(2) $\mathcal{B}$ <u>has the covering property</u> V_q, $\frac{1}{p} + \frac{1}{q} = 1$.

<u>Proof</u>

(1) $\Rightarrow$ (2). Associated to the basis $\mathcal{B}$ we can consider the maximal function $M_r (r > 0)$, defined on locally integrable functions f by the formula

$$M_r f(x) = \sup_{\substack{R \in \mathcal{B}(x) \\ \text{diameter}(R) \leq r}} \frac{1}{|R|} \int_R |f(y)| \, dy$$

According to III.3.4.(3), M_r is of weak type (p,p). Given a measurable bounded set E and given $\varepsilon > 0$, we pick an open set Ω s.t. $\Omega \supset E$ and $|\Omega - E| \leq \varepsilon$. From now on, we shall consider only the elements of the Vitali covering of E which are contained in Ω and have diameter less than r. Obviously they constitute another Vitali covering of E; we shall denote by V that covering.

Since the measures of the elements of V are bounded, we can choose an element R_1 such that $|R_1| \geq \frac{1}{2} \sup \{|R|, R \in V\}$.

Suppose that we have chosen $R_1, \ldots, R_k$. Then we divide the family V in two classes:

1) Elements R s.t. $|R \cap \bigcup_{j \leq k} R_j| \leq \frac{1}{2} |R|$.

2) Elements R s.t. $|R \cap \bigcup_{j \leq k} R_j| > \frac{1}{2} |R|$.

We eliminate the second class and observe that the first class constitutes a Vitali covering of $E - \bigcup_{j \leq k} R_j$.

Now we choose R_{k+1} to be an element of the first class such that $|R_{k+1}| \geq \frac{1}{2} \sup \{|R|; R \text{ is in the first class}\}$. By induction we get a sequence $\{R_k\}$ such that $|E_k| \geq \frac{1}{2} |R_k|$ where $E_k = R_k - \bigcup_{j<k} R_j$ and furthermore $|R_k|$ is of the order of the

biggest possible. From this, and using the fact that $\mathcal{B}$ differentiates integrals of functions in L^p, it is easy to see that $|E - \bigcup R_k| = 0$. The relation $|\bigcup R_k - E| \leq \varepsilon$ is an inmediate consequence of the fact that $R_k \subset \Omega$ for every k.

Next we consider the linear operator

$$Tf(x) = \sum \frac{1}{|R_k|} \int_{R_k} f(y)dy \cdot \chi_{E_k}(x)$$

and its formal adjoint

$$Sf(x) = \sum \frac{1}{|R_k|} \int_{E_k} f(y)dy \cdot \chi_{R_k}(x).$$

Observe that

$$|Tf(x)| \leq M_r f(x)$$

and

$$S(\chi_{\bigcup R_k}) \geq \frac{1}{2} \sum \chi_{R_k}.$$

Since M_r is of weak type (p,p), we have that the family of operators like T (corresponding to different sequences $\{R_k\}$) is a uniformly bounded family of linear operators from $L^p(\mathbb{R}^n)$ to the Lorentz space $L(p,\infty)$. Therefore their duals T* are uniformly bounded operators from $L(p,\infty)^*$ to L^q. But since $L(p,\infty)$ is the dual Banach space of $L(q,1)$ $(1/p + 1/q = 1)$, it follows that the operators S are uniformly bounded from the Lorentz space $L(q,1)$ to L^q. That is, there exists a constant C independent of E, ε and the sequence $\{R_k\}$, such that:

$$\|\sum \chi_{R_k}\|_q \leq \frac{1}{2} C \|\chi_{\bigcup R_k}\|^*_{q,1} \leq C|E|^{1/q}$$

(This is true because $||\chi_F||^*_{q,r} = |F|^{1/q}$ for every measurable set F, and every r $1 \leq r < \infty$, see R. Hunt, On L(p,q) spaces, L'Ens. Mathematique 12 (1.966), 249-76).

The implication (2) => (1) is straightforward.

q.e.d.

REMARK. The same linearization technique also allows us to prove the following two results:

1º) If $\mathcal{B}$ differentiates integrals of functions in L^1 then it has a covering property of exponential type i.e. there exists a constant $C > 0$ such that given a $\mathcal{B}$-Vitali covering of the set E, we can find a sub-covering $\{R_k\}$ satisfying

$$||\exp (C \sum \chi_{r_k}(x))||_1 \leq |E|$$

2º) If $\mathcal{B}$ differentiates integrals of functions in $L \log^+ L$ (for example the basis of intervals in $\mathbb{R}^2$), then there exists $C > 0$ such that, under the same conditions of 1), we have

$$||\exp (C \sum \chi_{R_k}(x))^{1/2}||_1 \leq |E|$$

However, these two covering properties are far from being the best possible for the corresponding situations.

The halo problem.

Let $\mathcal{B}$ be a differentiation basis in $\mathbb{R}^n$ (not necessarily invariant by translations) and let $\phi(u)$ be its halo function, that is

$$\phi(u) = \sup \{\frac{1}{|A|} |\{x : M\chi_A(x) > u^{-1}\}|, \text{ A bounded and with positive measure }\}, \; u \geq 1.$$

We can extend ϕ to $[0,\infty)$ by setting $\phi(u) = u$ for $u \in [0,1]$. Theorem 2 gives us an alternative proof of some results of Hayes and de Guzmán.

Theorem 2. Suppose that $\phi(u) = O(u^p)$ as $u \to \infty$ for some $1 \leq p < \infty$, then differentiates integrals of functions in $L_{loc}(p,1)$.

Proof. We shall show that $\mathcal{B}$ has the Vitali covering property V_q(weak) $1/p + 1/q = 1$. That is, there exists $C > 0$ such that given a bounded measurable set E, $\varepsilon > 0$ and a $\mathcal{B}$-Vitali covering of E, we can select a sequence $\{R_k\}$ satisfying $|\bigcup R_k \,\Delta\, E| \leq \varepsilon$ and

$$|\{x : \sum \chi_{R_k}(x) > \lambda\}| \leq C \frac{|E|}{\lambda^q} \quad \text{for every } \lambda > 0.$$

To see this we select a sequence $\{R_k\}$ as in Theorem 1 and we consider the linear operators T and T^*.

Then

$$|E_\lambda| = |\{x : \sum \chi_{R_k}(x) > \lambda\}| \leq$$

$$\leq \frac{2}{\lambda} \int_{E_\lambda} T^*\chi_E(x)dx = \frac{2}{\lambda} \int T\chi_{E_\lambda}(x)\chi_E(x)dx \leq$$

$$\leq \frac{2}{\lambda} ||\chi_E||_{q,1} ||T\chi_{E_\lambda}||_{p,\infty} \leq \frac{C^{1/q}}{\lambda} |E|^{1/q} |E_\lambda|^{1/p}$$

and therefore $|E_\lambda| \leq C \frac{|E|}{\lambda^q}$.

(The same argument shows that T^* is a bounded linear operator from $L(q,1)$ to $L(q,\infty)$.

The proof of the fact that V_q(weak) implies differentiation of integrals of functions in $L_{loc}(p,1)$ is straightforward.

q.e.d.

Corollary. If $\phi(u) = O(u^1)$ then $\mathcal{B}$ differentiates integrals of functions in $L^1(1 + \overset{+}{\mathrm{Log}}\, L)^1$).

Acknowledgment.

I wish to thank B. Rubio for having brought these problems to my attention and C. Fefferman for his helpful remarks while I was writing this paper.

APPENDIX II

A geometric proof of the strong maximal theorem

by Antonio Córdoba and Robert Fefferman

Consider in R^n the family B_n of rectangles with sides parallel to the coordinate axes. Associated with B_n, we can consider the maximal operator M_n defined on locally integrable functions f by the formula

$$M_n(f)(x) = \sup_{\substack{x \in R \\ R \in B_n}} \frac{1}{|R|} \int_R |f(y)| dy.$$

Then the strong maximal theorem can be stated in two equivalent ways:

(1) A quantitative form: M_n is bounded from the Orlicz space $L(1+(\log^+ L)^{n-1})$ to weak L^1, that is,

$$|\{x \mid M_n(f)(x) > \alpha\}| \leq A_n \int \frac{|f(x)|}{\alpha} (1 + (\log^+ \frac{|f(x)|}{\alpha})^{n-1}) dx$$

where A_n is some constant depending only on the dimension n, and $|E|$ denotes the Lebesgue measure of the set E.

(2) A qualitative form: The basis B_n differentiates integrals of functions which are locally in $L(\log^+ L)^{n-1}$, that is we have

$$\lim_{\substack{\mathrm{diam}(R) \to 0 \\ x \in R \in B_n}} \frac{1}{|R|} \int_R f(y)\, dy = f(x) \quad \text{for a.e. } x \in R^n$$

so long as f is locally in $L(\log^+ L)^{n-1}$.

This maximal theorem in its qualitative form is due to Jessen, Marcinkiewicz, and Zygmund [1935] and the basic idea of their proof is to dominate the operator M_n by the composition $M_{x_1} M_{x_2} \ldots M_{x_n}$ where M_{x_i} is the one dimensiona Hardy-Littlewood maximal operator in the direction of the ith coordinate axis.

The simplicity and elegance of this proof are obvious. On the other hand, it seemed desirable to have a geometric proof of the strong maximal theorem. The main reason is that we feel that in many cases, the only way to obtain results for operators intimately connected with the strong maximal function will be through a deep understanding of the geometry of rectangles. It is this understanding that we have done our best to achieve in this article.

Now, it was shown in Appendix I that under very general hypotheses, to study the properties of a maximal operator with respect to a family of bounded measurable sets is equivalent to studying covering properties of that family.

More precisely: Let $\mathcal{B}$ be a collection of bounded open sets in R_n, and let

$$M(f)(x) = \sup_{x \in R \in \mathcal{B}(x)} \frac{1}{|R|}\int_R |f(y)|\,dy \quad \text{if} \quad x \in \bigcup_{R \in \mathcal{B}} R$$

and $M(f)(x) = 0$ otherwise.

Definition. $\mathcal{B}$ has the covering property V_q, $1 \leq q \leq \infty$, if there exist constants $C < \infty$ and $c > 0$ so that given any subfamily $\{R_i\}_{i \in J}$ of we can find a sequence $\{\tilde{R}_k\} \subset \{R_i\}_{i \in J}$ satisfying the following conditions:

(1) $|\bigcup_k \tilde{R}_k| \geq c\,|\bigcup_{i \in J} R_i|$.

(2) $||\sum \chi_{\tilde{R}_k}||_q \leq C|\bigcup R_i|^{1/q}$.

Proposition 1. The maximal operator M is of weak type (p,p) if and only if $\mathcal{B}$ has the covering property V_q, $1/p + 1/q = 1$ $1 < p \leq \infty$.

Proof

(a) Suppose that $\mathcal{B}$ has the covering property V_q. Let $\alpha > 0$ and $E_\alpha = \{x : M(f)(x) > \alpha\}$.

Then $E_\alpha = \bigcup_{i \in J} R_i$ where $R_i \in \mathcal{B}$, and

$$\frac{1}{|R_i|}\int_{R_i}|f(y)|dy > \alpha, \text{ for every } i \in J.$$

Choose $\tilde{R}_k$ as in the definition of V_q. Then

$$|E_\alpha| = |\bigcup_{i\in J} R_i| \leq c^{-1}|\bigcup \tilde{R}_k| \leq c^{-1}\frac{1}{\alpha}\sum_k \int_{\tilde{R}_k}|f(y)|dy =$$

$$= c^{-1}\frac{1}{\alpha}\int \sum_k \chi_{\tilde{R}_k}(y)|f(y)|dy \leq Cc^{-1}\frac{1}{\alpha}|E_\alpha|^{1/q}||f||_p,$$

so

$$|E_\alpha| \leq (Cc^{-1})^p \frac{||f||_p^p}{\alpha^p}$$

which is the desired weak type (p,p) inequality.

(b) Conversely, suppose now that M is bounded from L^p to the Lorentz space $L(p,\infty)$, weak L^p. Given the family $\{R_i\}$ as in the statement of the Proposition, let us assume the existence of a subsequence $\{\tilde{R}_k\}$ such that $|\bigcup \tilde{R}_k| \geq c\,|\bigcup R_i|$ and with the disjoint part property P_1

$$P_1 : |\tilde{R}_k \cap \bigcup_{j<k} \tilde{R}_j| \leq \frac{1}{2}|\tilde{R}_k|.$$

Then we claim that the sequence $\tilde{R}_k$ also satisfies

$$||\sum \chi_{\tilde{R}_k}||_q \leq c|\bigcup R_i|^{1/q}.$$

To see this let $E_k = \tilde{R}_k - \bigcup_{j<k}\tilde{R}_j$, so that $|E_k| \geq \frac{1}{2}|\tilde{R}_k|$. Also define a linear operator T by

$$Tf(x) = \sum_k \left(\frac{1}{|\tilde{R}_k|}\int_{\tilde{R}_k} f(y)\,dy\right)\chi_{E_k}(x),$$

and notice that its adjoint is given by

$$T^*f(x) = \sum_k \left(\frac{1}{|\tilde{R}_k|}\int_{E_k} f(y)\,dy\right)\chi_{\tilde{R}_k}(x).$$

Observe that

$$|Tf(x)| \leq M(f)(x), \text{ and } T^*(\chi_{\cup\tilde{R}_k}) \geq \frac{1}{2}\sum\chi_{\tilde{R}_k}.$$

Because M is bounded from L^p to $L(p,\infty)$, it follows that T is bounded from L^p to $L(p,\infty)$ with a bound independent of $\tilde{R}_k$. Therefore T^* is bounded from $L(q,1)$ to L^q, so that

$$||\sum\chi_{R_k}||_q \leq C|\cup\tilde{R}_k|^{1/q}$$

(C independent of $\tilde{R}_k$).

Let us show that we can select the sequence $\tilde{R}_k$ with the required sparseness property P_1. First of all we can assume that $\{R_i\}$ is a sequence $R_1, R_2, \ldots, R_j, \ldots$. Take $\tilde{R}_1 = R_1$, and suppose that we have chosen $\tilde{R}_1, \ldots, \tilde{R}_k$. Then $\tilde{R}_{k+1}$ is the first rectangle in the sequence $\{R_j\}$ after $\tilde{R}_k$ with the following property

$$|\tilde{R}_{k+1} \cap \bigcup_{j\leq k} \tilde{R}_j| \leq \frac{1}{2}|\tilde{R}_{k+1}|.$$

And observe that $\bigcup R_i \subset \{x \mid M(\chi_{\bigcup_k \tilde{R}_k})(x) \geq \frac{1}{2}\}$.

<u>REMARK</u>. The same methods as those used in the preceding proof also yield the following results:

(1) If M is of weak type (1,1) and if $\{R_k\}$ has the disjointness property P_1,

then

$$||\exp(\sum \chi_{R_k})||_1 \leq C|\bigcup R_k|,$$

for some absolute constant $C < \infty$.

(2) If M is bounded from $L(1 + (\log^+ L)^r)$ to weak L^1, and if $\{R_k\}$ has property P_1, then

$$||\exp(\sum \chi_{R_k})^{1/(r+1)}||_1 \leq C|\bigcup R_k|$$

for some absolute constant $C < \infty$.

We are now ready to prove the main result of this paper.

Theorem 2. Suppose that $\{R_i\}_{i\in J}$ is a family of rectangles in R^n $(n > 1)$ with sides parallel to the axes. Then there is a sequence $\{\tilde{R}_k\} \subset \{R_i\}_{i\in J}$ satisfying

(i) $|\bigcup \tilde{R}_k| \geq c_n |\bigcup_{i\in J} R_i|$

and

(ii) $||\exp(\sum \chi_{\tilde{R}_k})^{\frac{1}{n-1}}||_1 \leq C_n |\bigcup_{i\in J} R_i|$

where $c_n > 0$ and $C_n < \infty$ are constants depending only on the dimension n.

Proof. The proof is by induction on n. For $n \geq 3$, let us consider the passage from n-1 to n. Assume the result is true for n-1. Then the strong maximal theorem is also true in R^{n-1}, and therefore using the remark of proposition 1, we know that if $\{T_k\}$ is a sequence of rectangles in R^{n-1} with the disjoint parts property P_1, then we have the estimate

$$||\exp(\sum \chi_{T_k})^{\frac{1}{n-1}}||_1 \leq C_{n-1}|\bigcup T_k|.$$

Starting with a family $\{R_i\}_{i\in J}$ in R^n, (which we assume to be a sequence), we claim the existence of a subfamily $\{\tilde{R}_k\}$ such that

$$|\bigcup \tilde{R}_k| \geq c_n \, |\bigcup_{i\in J} R_i|$$

and with the sparseness property P_2:

$$P_2 : |\tilde{R}_k \cap (\bigcup_{j\neq k} \tilde{R}_j)| < \frac{1}{2} |\tilde{R}_k| .$$

To see this we choose first (assuming $|R_i|$ decreases as i increases) a subsequence $R_1, R_2, \ldots, R_k, \ldots$ from the family $\{R_i\}_{i\in J}$ as in proposition 1, and observe that

$$\bigcup_{i\in J} R_i \subset \bigcup_{j=1}^{n} \{x \mid M_{n-1}^{x_j}(\chi_{\bigcup R_k^*})(x) > \frac{1}{2n}\}$$

where $M_{n-1}^{x_j}$ denotes the strong maximal function in hyperplanes perpendicular to the direction x_j, and R_k^* is the double of R_k.

Therefore

$$|\bigcup_{i\in J} R_i| \leq 2^{n+1} n(\log 2n)^{n-1} \, c_{n-1} |\bigcup_{k=1}^{\infty} R_k| .$$

Using this, we can select a finite sequence $R_1, R_2, \ldots, R_M$ so that

$$|\bigcup_{k=1}^{M} R_k| \geq \frac{1}{2} |\bigcup_{k=1}^{\infty} R_k| .$$

Writing this finite sequence backwards, i.e., $R_M, R_{M-1}, \ldots, R_1$, and applying the method of proposition 1 to this inverted sequence, we arrive at a subsequence $\{\tilde{R}_k\}_{k=1,2,\ldots,N}$ satisfying P_2 and

$$|\bigcup_{k=1}^{N} \tilde{R}_k| \geq c_n |\bigcup_{i\in J} R_i|$$

where

$$c_n = |2n\ (\log 2n)^{n-1}\ 2^{n+1}\ C_{n-1}|^{-2}.$$

We now order the n dimensional rectangles $\tilde{R}_k$ so that their side lengths in the x_n direction are decreasing. The point is that when we slice the family $\{\tilde{R}_k\}$ with a hyperplane perpendicular to the x_n axis, the resulting n-1 dimensional rectangles must satisfy the disjoint parts condition P_1. Therefore for each fixed x_n,

$$\int \exp\ [(\sum \chi_{T_k}(x_1,x_2,\ldots,x_n))^{\frac{1}{n-1}}]\ dx_1,dx_2,\ldots,dx_{n-1} \leq C_{n-1}|\ \ T_k|.$$

Integrating in the x_n variable we obtain the desired passage from n-1 to n. To get the case n = 2, we have only to observe that if I_1, I_2, ..., I_k, ... are dyadic intervals satisfying condition P_1 then

$$||\exp\ (\sum \chi_{I_k})||_1 \leq C|\bigcup I_k|,$$

which follows from the method of proof of proposition 1. Then the reduction step applies to finish the proof.

Of course, as in the case of L^p, in the first proposition the estimate

$$||\exp\ (\sum \chi_{R_k})^{\frac{1}{n-1}}||_1 \leq C|\ \ R_k|$$

which is satisfied by all sequences of R_k in the space R^n satisfying

$$|R_k \cap (\bigcup_{j\neq k} R_j)| < \frac{1}{2}\ |R_k|,$$

proves the strong maximal theorem.

Now, let us notice one point. We have considered so far, two different sparseness conditions on rectangles:

$$P_1 : |R_k \cap (\bigcup_{j<k} R_j)| < \frac{1}{2} |R_k|$$

and

$$P_2 : |R_k \cap (\bigcup_{j\neq k} R_j)| < \frac{1}{2} |R_k| .$$

Although, off hand, these two conditions appear more or less the same, note that by the reduction argument of the proof above, property P_1 cannot imply anything more than

$$||\exp (\sum \chi_{R_k})^{1/n}||_1 \leq c_n |\bigcup R_k| ,$$

lest we improve upon the class $L[(\log^+ L)^{n-1} + 1]$ in the statement of the strong maximal theorem. However the stronger P_2 yields

$$||\exp (\sum \chi_{R_k})^{\frac{1}{n-1}}||_1 \leq c_n |\bigcup R_k| .$$

We would now like to present another proof of the strong maximal theorem for the case n = 2. First, it is somewhat more geometric than the proof just given, and second, it served as motivation for considering the stronger sparseness condition P_2.

So, given rectangles $\{R_j\}_{j=1,2,...}$ with sides parallel to the axes in the plane we shall show directly that if

$$|R_k \cap (\bigcup_{j \neq k} R_j)| < \frac{1}{2} |R_k|,$$

then $|\{x \mid x \text{ is in at least } r \text{ distinct } R_j^! x\}| \leq C2^{-(r-1)} |\bigcup R_k|$, which gives

$$||\exp (\sum \chi_{R_k})||_1 \leq C_1 |\bigcup R_k|.$$

To get this, assume that the R_k are dyadic and have their longest side parallel to the x-axis. Introducing the eccentricity

$$e(R_k) = \frac{\text{length of the side of } R_k \text{ along the x-axis}}{\text{length of the side of } R_k \text{ along the y-axis}}$$

partially order the R_k by saying that $R_k \geq R_j$ if $R_k \cap R_j \neq \phi$ and $e(R_k) \leq e(R_j)$.

Take any fixed $k \geq 1$, and consider the rectangle R_k. We shall show that the measure of the union U_k of all sets of the form $R_k \cap S_1 \cap S_2 \cap \ldots S_{r-1} \cap S_r$, (here the S_i are rectangles in the sequence $R_1, R_2, \ldots$ which are $\leq R_k$) has measure $\leq 2^{-r}|R_k|$. In fact we can argue as follows: if we project all rectangles $\leq R_k$ on the x and y axes we shall get

$$S_i = S_i^x \times S_i^y,$$

$$R_k = R_k^x \times R_k^y$$

(where S_i^x and S_i^y are dyadic intervals on the line) and because $S_i \leq R_k$, we have $S_i^x \supset \supset R_k^x$ for each i. Thus we have found the projection U_k^x of U_k on the x axis to be R_k^x. And by the sparseness property of the R_j, U_k^y is the union of all r+1-fold intersections of dyadic intervals contained in R_k^y having the property that

$$|D_i \cap (\bigcup_{D_j \subsetneq D_i} D_j)| < \frac{1}{2} |D_i|$$

It is then the case that

$$|U_k^y| \leq 2^{-r} |R_k^y|,$$

and we have $|U_k| \leq 2^{-r}|R_k|$. Since $\{x \mid x$ is contained in at least r+1 distinct $R_i\} \subset \subset \bigcup_{k \geq 1} U_k$ we have

$$|\{x \mid x \text{ is contained in at least r+1 distinct } R_i\}| \leq$$

$$\leq 2^{-r}\sum | \; R_k| \leq 2^{-r} \, c|\bigcup R_k|.$$

APPENDIX III

Equivalence between the regularity property and the differentiation of L^1 for a homothecy invariant basis.

by Roberto Moriyón

This appendix shows that for a B - F basis in $\mathbb{R}^n$ that is invariant by homothecies, it is equivalent to differentiate L^1 and to be regular with respect to cubic intervals. This result is interesting in view of the example of Remark (8) of I.3. At the same time we obtain an affirmative answer to the halo conjecture of Chapter VIII in the case where $\psi(u)$, the halo function, behaves at infinity like u.

THEOREM. Let $\mathcal{B}$ be a B - F basis that is homothecy invariant. Let

$$K = \bigcup\{R \in \mathcal{B}(0) : |R| \leq 1\}.$$

Then the following statements are equivalent:

a) $\mathcal{B}$ differentiates L^1.

b) The maximal operator M of $\mathcal{B}$ is of weak type (1,1).

c) If $\psi : [0,\infty) \to [0,\infty)$ is the halo function of $\mathcal{B}$, then there exist two constants $c_1, c_2 > 0$ such that $c_1 u \leq \psi(u) \leq c_2 u$.

d) K is of finite measure.

e) K is bounded.

f) $\mathcal{B}$ is regular with respect to $\mathcal{B}_1$, the basis of cubic intervals.

Finally, if properties a) - d) do not hold, then the set of all functions f of L^1 such that there is at least a point x_f so that $\mathcal{B}$ differentiates $\int f$ at x_f is of the first category in L^1.

Proof. We shall prove a) => b) => c) => d) => e) => f) => a).

a) => b). This implication has already been proved in III.3.4.

b) => c). Recall the definition of the halo function. For $u > 1$

$$\psi(u) = \sup_{|A|>0} \frac{|\{M\chi_A > \frac{1}{u}\}|}{|A|}$$

Since b) implies

$$|\{M\chi_A > \frac{1}{u}\}| \leq c_2 u \, |A|$$

we get $\psi(u) \leq c_2 u$.

In order to prove $c_1 u \leq \psi(u)$, take $R \in \mathcal{B}$ and let $\mathcal{B}_0$ be the B - F basis invariant by homothecies generated by R. Since $\mathcal{B}_0$ is regular with respect to $\mathcal{B}_1$ we easily see that the halo function ψ_0 of $\mathcal{B}_0$ behaves like ψ_1, that of $\mathcal{B}_1$. But for ψ_1 we already know $\psi_1(u) \geq c_1^* u$, and so $\psi_0(u) \geq c_1 u$. Since ψ majorizes ψ_0, we get $\psi(u) \geq c_1 u$.

c) => d). The set K is clearly open and so measurable. Let, for $k = 1,2,\ldots$, $B_k = B(0,\frac{1}{k})$ and

$$K_k = \bigcup \{R \in \mathcal{B}(0) : |R| \leq 1, R \supset B_k\}$$

Then $\{K_k\}$ is an expanding sequence of open sets and $K = \bigcup_1^\infty K_k$. Furthermore

$$K_k \subset \{M\chi_{B_k} > \frac{1}{2} |B_k|\}$$

Hence, by c), for each k

$$|K_k| \leq \psi(\frac{2}{|B_k|}) \, |B_k| \leq 2c_2$$

and so $|K| < \infty$.

d) $\Rightarrow$ e). Assume that K is not bounded. Since $\mathcal{B}$ is homothecy invariant, there exists a sequence $\{R_k\} \subset \mathcal{B}(0)$ such that $|R_k| = 1$ and $\delta(R_k) \to \infty$. We now define a new sequence $\{S_k\} \subset \mathcal{B}(0)$, $|S_k| = 1$ in the following way. Take $S_1 = R_1$. Consider now R_2. Two cases may occur. If $|R_2 \cap S_1| \leq \frac{1}{2}$, then $S_2 = R_2$. In case $|R_2 \cap S_1| > \frac{1}{2}$ and there is some point $x_2 \in R_2$ such that $|x_2| > 4\delta(S_1)$, then we set $S_2 = -x_2 + R_2$. If $|R_2 \cap S_2| > \frac{1}{2}$ but there is no such point $x_2 \in R_2$ with $|x_2| > 4\delta(S_1)$ we advance in the sequence $\{R_k\}$ until we find a first set R_{k_2} such that either $|R_{k_2} \cap S_1| \leq \frac{1}{2}$ (and then we set $S_2 = R_{k_2}$) or $|R_{k_2} \cap S_1| > \frac{1}{2}$ and there is $x_{k_2} \in R_{k_2}$ such that $|x_{k_2}| > 4\delta(S_1)$ (and then we set $S_2 = -x_{k_2} + R_{k_2}$). In order to choose S_3 we advance further in the sequence $\{R_k\}$ until we find a first set R_{k_3} such that either

$$|R_{k_3} \cap (\bigcup_1^2 S_j)| \leq \frac{1}{2}$$

(and then we set $S_3 = R_{k_3}$) or

$$|R_{k_3} \cap (\bigcup_1^2 S_j)| > \frac{1}{2}$$

and there is $x_{k_3} \in R_{k_3}$ such that

$$|x_{k_3}| > 4\, \delta(\bigcup_1^2 S_j).$$

And so on. Observe now that, by construction, for each $k \geq 2$

$$|S_k - \bigcup_{j=1}^{k-1} S_j| > \frac{1}{2}$$

and the sets

$$L_k = S_k - \bigcup_{j=1}^{k-1} S_j$$

are pairwise disjoint. Since $K \supset \bigcup L_k$ we get $|K| = \infty$.

e) => f). If K is bounded, there is a cubic interval E containing K. Let $R \in \mathcal{B}$. Let $R_0 \in$ (0) be homothetic to R and $|R_0| = 1$. Let E* be the cubic interval obtained from E by the same homothecy that transforms R_0 into R. Then $E^* \supset R$ and

$$\frac{|E^*|}{|R|} = \frac{|E|}{|R_0|} = |E|$$

which is independent of R. Hence $\mathcal{B}$ is regular with respect to $\mathcal{B}_1$.

In order to prove the last part of the theorem it is enough to observe that the technique used in IV. 2.1. for the proof of Sak's rarity theorem can be applied here without any substantial change.

REMARKS 1. The equivalence of b) and c) gives a positive answer to the halo conjecture of Chapter 8 when $\psi(u)$ behaves like u.

REMARK 2. The statements a) - f) of the theorem are also equivalent to the following condition:

g) There exists c, $0 < c \leq 1$ such that for each $\lambda > 1$, for each measurable set A and for each Vitali cover $\mathcal{V}$ of A by sets of $\mathcal{B}$ there exists a disjoint sequence $\{R_k\} \subset \mathcal{V}$ such that

$$|A - \bigcup R_k| \leq \lambda(1-c)\ |A|$$

Furthermore, in case g) holds, the best possible constant c satisfying the above inequality is

$$c_0 = \inf \{\frac{|B|}{|\overline{B}|} : B \in \mathcal{B}\}$$

Hence if for each $B \in \mathcal{B}$, we have $|B| = |\overline{B}|$, then $c_0 = 1$ and so $\mathcal{B}$ has the

Vitali property.

The proof of this property, a little technical, but not difficult, will be ommitted.

APPENDIX IV

On the derivation properties of a class of bases

by Roberto Moriyón

In this note we present the solution to the problem proposed in VI.4. and we state a general theorem that can be proved by the same type of argument used here.

We consider the translation invariant B - F basis $\mathcal{B}$ constituted by all open intervals in R^2 such that if d is the length of the smaller side and D is the one of the bigger side we have $D^2 \leq d \leq D \leq 1$.

THEOREM 1. The basis $\mathcal{B}$ differentiates $L(1 + \log^+ L)$ and this is the best space it differentiates in the following sense: If $\Phi : [0,\infty) \to [0,\infty)$, $\Phi(0) = 0$ is an increasing continuous function and if $\mathcal{B}$ differentiates $\Phi(L)$, then any function in $\Phi(L)$ is locally in $L(1 + \log^+ L)$.

Proof. That $\mathcal{B}$ differentiates $L(1 + \log^+ L)$ is obvious, since $\mathcal{B}$ is a subbasis of $\mathcal{B}_2$, the basis of all intervals. In order to prove the second part of the theorem we proceed as follows.

For each $x \in (0,1)$, Q_x will denote the interval of R^2 $Q_x = (0,x^{3/2}) \times (0,x^2)$ and c_x will be the equilateral hyperbola passing through (x,x^2), as in the following picture

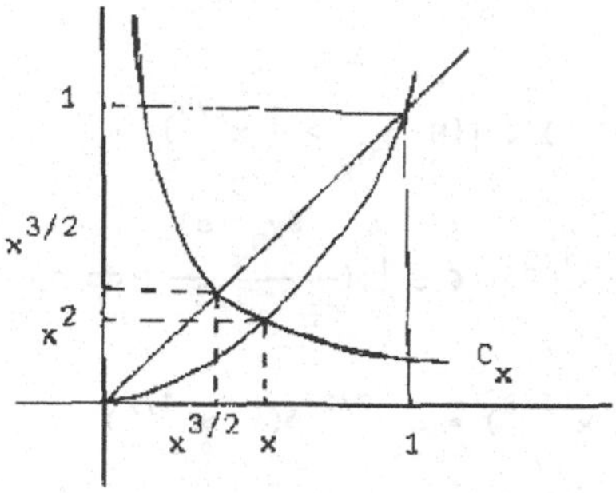

Fig. 24

Clearly, each rectangle of $\mathcal{B}$ with a vertex at (0,0) and the opposite one on c_x contains Q_x and have the same measure, i.e. if $\theta \in [x^{3/2},x]$, $Q_{x,\theta} = (0,\theta) \times (0,x^3,\theta^{-1})$ then $|Q_{x,\theta}| = x^3$. Also $\delta(Q_{x,\theta}) \leq 2r$. Hence, since

$$\frac{|Q_{x,\theta} \cap Q_x|}{|Q_{x,\theta}|} = \frac{|Q_x|}{|Q_{x,\theta}|} = \frac{x^{7/2}}{x^3} = x^{1/2}$$

we get for each $x \in (0,1)$ and each $\theta \in [x^{3/2},x]$,

$$Q_{x,\theta} \quad \{y\in R^2 : M_{2x}\, \chi_{Q_x}(y) > \frac{1}{2}\, x^{1/2}\}$$

and so

$$|\{y\in R^2 : M_{2x}\chi_{Q_x}(y) > \frac{1}{2}\, x^{1/2}\}| \geq$$

$$\geq |\bigcup_{x^{3/2} \leq \theta \leq x} Q_{x,\theta}| = x^3 + \int_{x^{3/2}}^{x} \frac{x^3 ds}{s} = x^3(1 + \log x^{-1/2}) \qquad (*)$$

If $\mathcal{B}$ differentiates $\Phi(L)$, according to the Remark (2) of III.3., there exist $r > 0$ and $c > 0$ such that for each measurable function f and for each $\lambda > 0$

$$|\{y\in R^2 : M_r f(y) > \lambda\}| \leq c \int \Phi(\frac{8|f(s)|}{\lambda})ds \qquad (**)$$

Hence, if $x \leq x_0 = \min(1,\frac{r}{2})$, having into account (*) and (**)

$$x^3(1 + \log x^{-1/2}) \leq |\{M_{2x}\chi_{Q_x} > \frac{1}{2}\, x^{1/2}\}| \leq$$

$$\leq |\{M_r\chi_{Q_x} > \frac{1}{2}\, x^{1/2}\}| \leq c \int \Phi(\frac{8\chi_{Q_x}(s)}{\frac{1}{2}\; x^{1/2}})\; ds =$$

$$= c\, |Q_x|\, \Phi(16\, x^{-1/2}) = c\, x^{7/2}\, \Phi(16\, x^{-1/2})$$

If we set $16\, x^{-1/2} = \lambda$, then we get for $\lambda \geq \lambda_0 = 16\, x_0^{-1/2}$,

$$\Phi(\lambda) \geq k\lambda(1 + \log \lambda)$$

and this obviously implies the statement of the theorem.

In order to state the general theorem announced at the beginning we shall introduce some notation.

For a family $\mathcal{A}$ of open bounded sets in R^2 we shall call $\mathcal{B}(\mathcal{A})$ the translation invariant B - F basis generated by $\mathcal{A}$.

If $x,y > 0$ we shall call

$$I(x,y) = (0,x) \times (0,y).$$

Let f, g be two functions from $(0,\varepsilon)$ to $(0,\infty)$ such that

$$\lim_{x\to 0} f(x) = \lim_{x\to 0} g(x) = 0$$

and $f(x) \leq g(x)$ for each $x \in (0,\varepsilon)$.

We shall call

$$\mathcal{A}_f = \{I(x,f(x)) : x \in (0,\varepsilon)\}$$

$$\mathcal{A}_{f,g} = \{I(x,y) : x \in (0,\varepsilon),\ f(x) \leq y \leq g(x)\}$$

For two translation invariant B - F bases $\mathcal{B}$ and $\mathcal{B}'$ we shall say that $\mathcal{B}$ is locally regular with respect to $\mathcal{B}'$ whenever there exist $r > 0$ and $c > 0$ such that for each $R \in \mathcal{B}$ with $\delta(R) \leq r$, there exists $R' \in \mathcal{B}'$ such that $R \subset R'$ and $|R'| \leq c|R|$.

<u>THEOREM</u> 2. <u>Let</u> f, g <u>be two functions from</u> $(0,\varepsilon)$ <u>to</u> $(0,\infty)$ <u>as above and such that the functions</u> $h(x) = xf(x)$, $j(x) = xg(x)$ <u>are continuous and strictly increasing.</u>

<u>Then, if</u> $\mathcal{B}(\mathcal{A}_{f,g})$ <u>differentiates some space</u> $\Phi(L)$ (<u>as in Theorem</u> 1) <u>strictly greater than</u> $L(1 + \log^+ L)$, <u>there exists</u> $c > 0$ <u>such that</u> $\mathcal{B}(\mathcal{A}_{f,g})$ <u>is locally regular with</u> respect to $\mathcal{B}(\mathcal{A}_{cf})$.

BIBLIOGRAPHY

The following list contains merely the works referred to in the text and is rather incomplete. Anyone interested in the various aspects of the differentiation theory of integrals and measures should consult the following two works. Both contain very many additional references.

Bruckner, A. M., Differentiation of integrals, Amer. Math. Monthly 78 (1971) (Slaught Memorial Paper no. 12).

Hayes, C. A. and Pauc, C. Y., Derivation and martingales (Springer, Berlin, 1970).

Alfsen, E.M. [1965]

Some coverings of Vitali type, Math. Ann. 159 (1965), 203-216.

Banach, S. [1924]

Sur un theorème de M. Vitali, Fund. Math. 5(1924), 130-136.

Besicovitch, A.S. [1928]

On Kakeya's problem and a similar one, Math. Z. 27(1928), 312-320.

Besicovitch, A.S. [1935]

On differentiation of Lebesgue double integrals, Fund. Math. 5(1935), 209-216.

Besicovitch, A.S. [1945]

A general form of the covering principle and relative differentiation of additive functions, Proc.Cambridge Philos.Soc. 41(1945), 103-110.

Besicovitch, A.S. [1946]

A general form of the covering principle and relative differentiation of additive functions, Proc. Cambridge Philos. Soc. 42(1946), 1-10.

Boas, R.P. [1960]

A primer of real functions (Math. Assoc. Amer. 1960).

Burkill, J.C. [1951]

On the differentiability of multiple integrals, J. London Math. Soc. 26 (1951), 244-249.

Busemann, H. and Feller, W. [1934]

Zur Differentiation der Lebesgueschen Integrale, Fund. Math. 22(1934), 226--256.

Calderón, A.P. and Zygmund, A. [1956]

On singular integrals, Amer. J. Math. 18(1956), 289-309.

Calderón, A.P. and Zygmund, A. [1952]

On the existence of certain singular integrals, Acta Math. 88 (1952), 85-139.

Calderón, C.P. [1973]

Differentiation through starlike sets in R^m, Studia Math. 48 (1973), 1-13.

Caratheodory, C. [1927]

Vorlesungen über reelle Funktionen[2] (Leipzig, 1927).

Coifman, R. and de Guzmán, M. [1970]

Singular integrals and multipliers on homogeneous spaces, Rev. Un. Mat. Argentina, 25 (1970), 137-143.

Coifman, R. and Weiss, Guido, [1971]

Analyse harmonique non-commutative sur certains espaces homogènes (Lecture Notes 242, Berlin, 1971)

Córdoba, A. and Gallego, A. [1970]

Una propiedad de conjuntos convexos, XI Reunión Mat. Españoles (Murcia, 1970)

Cotlar, M. [1959]

Condiciones de continuidad de operadores potenciales y de Hilbert (Univ. de Buenos Aires, 1959).

Cunningham, F. [1971]

The Kakeya problem for simply connected and for star-shaped sets, Amer. Math. Monthly 78(1971), 114-129.

Cunningham, F. [1974]

Three Kakeya problems, Amer. Math. Monthly 81(1.974), 582-592.

Davies, R.O. [1953]

Accessibility of plane sets and differentiation of functions of two variables (Ph. D. Dissertation, Cambridge Univ., 1953).

Ellis, H.W. and Jeffery, R.L. [1967]

Derivatives and integrals with respect to a base function of generalized bounded variation, Canad. J. Math. 19(1967), 225-241.

Fefferman, C. [1971]

The multiplier problem for the ball, Ann. of Math. 94(1971), 330-336.

de Guzmán, M. [1970]

Singular integrals with generalized homogeneity, Rev. Acad. Ci. Madrid 64 (1970), 77-137.

de Guzmán, M. [1970]*

A covering lemma with applications to differentiability of measures and singular integral operators, Studia Math. 34 (1970), 299-317.

de Guzmán, M. [1972]

On the derivation and covering properties of a differentiation basis, Studia Math. 44(1972), 359-364.

de Guzmán, M. [1972]*

An extension of Sard's theorem, Bol. Soc. Brasileira Mat. 3(1972), 133-136.

de Guzmán, M. [1973]

On the halo problem in differentiation theory (to be published).

de Guzmán, M. [1974]

An inequality for the Hardy-Littlewood maximal operator with respect to a product of differentiation bases, Studia Math. 49(1972), 185-194.

de Guzmán, M. [1974]*

A general form of the Vitali lemma. (to be published).

de Guzmán, M. and Welland, G.V. [1971]

On the differentiation of integrals, Rev. Un. Mat. Argentina 25(1971), 253-276.

Hardy, G.H. and Littlewood, J.E. [1930]

A maximal theorem with function-theoretic applications, Acta Math. 54 (1930), 81-116.

Hayes, C.A. [1952]

Differentiation with respect to ϕ-pseudo-strong blankets and related problems, Proc. Amer. Math. Soc. 3(1952), 283-296.

Hayes, C.A. [1952]*

Differentiation of some classes of set functions, Proc. Cambridge Philos. Soc. 48(1952), 374-382.

Hayes, C.A. [1958]

A sufficient condition for the differentiation of certain classes of set functions, Proc. Cambridge Philos. Soc. 54(1958), 346-353.

Hayes, C.A. [1966]

A condition of halo type for the differentiation of classes of integrals, Canad. J. Math. 18(1966), 1015-1023.

Hayes, C.A. [1975]

A necessary and sufficient condition for the derivation of some classes of set functions. (to be published).

Hayes, C.A. and Pauc, C.Y. [1955]

Full individual and class differentiation theorems in their relations to halo and Vitali properties, Canad. J. Math. 7(1955), 221-274.

Herz, C. [1968]

The Hardy-Littlewood maximal theorem (Warwick University, Warwick 1968).

Iseki, K. [1960]

On the covering theorem of Vitali, Proc. Japan Acad. 36 (1960), 630-635.

Jeffery, R.L. [1932]

Non-absolutely covergent integrals with respect to functions of bounded variation, Trans. Amer. Math. Soc. 34 (1932), 645-675.

Jeffery, R.L. [1958]

Generalized integrals with respect to functions of bounded variation, Canad. J. Math. 10(1958), 617-626.

Jessen, B., Marcinkiewicz, J. and Zygmund, A. [1935]

Note on the differentiability of multiple integrals, Fund, Math. 25(1935), 217-234.

John, F. [1948]

Extremum problems with subsidiary conditions, in Studies and Essays (Interscience, New York, 1948).

Kakeya, S. [1917]

Some problems on maxima and minima regarding ovals, Tôhoku Sci. Reports 6 (1917), 71-88.

Lebesgue, H. [1910]

Sur l'integration des fonctions discontinues, Ann. Ecole. Norm. 27(1910), 361-450.

Morse, A.P. [1947]

Perfect blankets, Trans. Amer. Math. Soc. 6(1947), 418-442.

Nikodym, O. [1927]

Sur la mesure des ensembles plans dont tous les points sont rectilinéairement accessibles, Fund. Math. 10(1927), 116-168.

Papoulis, A. [1950]

On the strong differentiation of the indefinite integral, Trans. Amer. Math. Soc. 69(1950), 130-141.

Peral, I. [1974]

Nuevos métodos en diferenciación. (Tesis doctoral, Univ. de Madrid, 1974).

Perron, O. [1928]

Ueber einen Satz von Besicovitch, Math. Z. 28(1928), 383-386.

de Possel, R. [1936]

Sur la dérivation abstraite des fonctions d'ensemble, J. Math. Pures Appl. 15(1936), 391-409.

Rademacher, H. [1962]

A new construction of the Perron tree, in Studies in Mathematical Analysis (edited by Gilbarg, Solomon and others) (Stanford, 1962).

Riesz, F. [1932]

Sur un théorème de maximum de MM. Hardy and Littlewood, J. London Math. Soc. 7(1932), 10-13.

Riesz, F. [1934]

Sur les points de densité au sens fort, Fund. Math. 22(1934), 221-265.

Rubio, B. [1972]

Propiedades de derivación y el operador maximal de Hardy-Littlewood (Tesis Doctoral, Univ. Madrid. 1971).

Saks, S. [1934]

Remarks on the differentiability of the Lebesgue indefinite integral, Fund. Math. 22(1934), 257-261.

Saks, S. [1933]

Theorie de l'Integrale, Monografie Matematyczne, Volume II, (Warszawa, 1933).

Schoenberg, I.J. [1962]

On the Besicovitch-Perron solution of the Kakeya problem, in Studies in Mathematical Analysis. (edited by Gilbarg, Solomon and others) (Stanford, 1962).

Stein, E.M. [1969]

Note on the class Llog L, Studia Math. 31 (1969), 305-310.

Stein, E.M. and Weiss, N.J. [1969]

On the convergence of Poisson integrals, Trans. Amer. Math. Soc. 140(1969), 34-54.

Vitali, G. [1908]

Sui gruppi di punti e sulle funzioni di variabili reali, Atti Accad. Sci. Torino, 43(1908), 75-92.

Whitney, H. [1934]

Analytic extensions of differentiable functions defined on closed sets, Trans. Amer. Math. Soc. 36(1934), 63-89.

Yano, S. [1951]

An extrapolation theorem, J. Math. Soc. Japan 3(1951), 296-305.

Zygmund, A. [1929]

Sur les fonctions conjuguées, Fund. Math. 13(1920), 284-303.

Zygmund, A. [1934]

On the differentiability of multiple integrals, Fund. Math. 23(1934), 143--149.

Zygmund, A. [1959]

Trigonometric series (Cambridge, 1959).

Zygmund, A. [1967]

A note on the differentiability of multiple integrals, Colloq. Math. 16(1967), 199-204.

A LIST OF SUGGESTED PROBLEMS

INDEX

Vol. 309: D. H. Sattinger, Topics in Stability and Bifurcation Theory. VI, 190 pages. 1973. DM 20,-

Vol. 310: B Iversen, Generic Local Structure of the Morphisms in Commutative Algebra. IV, 108 pages. 1973. DM 18,-

Vol. 311: Conference on Commutative Algebra. Edited by J. W. Brewer and E. A. Rutter. VII, 251 pages. 1973. DM 24,-

Vol. 312: Symposium on Ordinary Differential Equations. Edited by W. A. Harris, Jr. and Y. Sibuya. VIII, 204 pages. 1973. DM 22,-

Vol. 313: K. Jörgens and J. Weidmann, Spectral Properties of Hamiltonian Operators. III, 140 pages. 1973. DM 16,-

Vol 314: M. Deuring, Lectures on the Theory of Algebraic Functions of One Variable. VI, 151 pages. 1973. DM 16,-

Vol. 315: K. Bichteler, Integration Theory (with Special Attention to Vector Measures). VI, 357 pages. 1973. DM 29,-

Vol. 316: Symposium on Non-Well-Posed Problems and Logarithmic Convexity. Edited by R. J. Knops V, 176 pages. 1973. DM 20,-

Vol. 317: Séminaire Bourbaki - vol. 1971/72. Exposés 400-417. IV, 361 pages. 1973. DM 29,-

Vol. 318: Recent Advances in Topological Dynamics. Edited by A. Beck. VIII, 285 pages. 1973. DM 27,-

Vol. 319: Conference on Group Theory Edited by R. W. Gatterdam and K. W. Weston. V, 188 pages. 1973. DM 20,-

Vol. 320: Modular Functions of One Variable I. Edited by W. Kuyk V, 195 pages. 1973. DM 20,-

Vol. 321: Séminaire de Probabilités VII. Edité par P. A. Meyer. VI, 322 pages. 1973. DM 29,-

Vol. 322: Nonlinear Problems in the Physical Sciences and Biology. Edited by I. Stakgold, D. D. Joseph and D. H. Sattinger. VIII, 357 pages. 1973. DM 29,-

Vol. 323: J. L. Lions, Perturbations Singulières dans les Problèmes aux Limites et en Contrôle Optimal. XII, 645 pages. 1973. DM 46,-

Vol. 324: K. Kreith, Oscillation Theory. VI, 109 pages. 1973 DM 18,-

Vol. 325: C.-C. Chou, La Transformation de Fourier Complexe et L'Equation de Convolution. IX, 137 pages. 1973. DM 18,-

Vol. 326: A. Robert, Elliptic Curves. VIII, 264 pages. 1973. DM 24,-

Vol. 327: E. Matlis, One-Dimensional Cohen-Macaulay Rings. XII, 157 pages. 1973. DM 20,-

Vol. 328: J. R. Büchi and D. Siefkes, The Monadic Second Order Theory of All Countable Ordinals. VI, 217 pages. 1973. DM 22,-

Vol. 329: W. Trebels, Multipliers for (C, α)-Bounded Fourier Expansions in Banach Spaces and Approximation Theory. VII, 103 pages. 1973. DM 18,-

Vol. 330: Proceedings of the Second Japan-USSR Symposium on Probability Theory. Edited by G. Maruyama and Yu. V. Prokhorov VI, 550 pages. 1973. DM 40,-

Vol. 331: Summer School on Topological Vector Spaces. Edited by L. Waelbroeck. VI, 226 pages. 1973. DM 22,-

Vol. 332: Séminaire Pierre Lelong (Analyse) Année 1971-1972. V, 131 pages. 1973 DM 18,-

Vol. 333: Numerische, insbesondere approximationstheoretische Behandlung von Funktionalgleichungen. Herausgegeben von R. Ansorge und W. Törnig. VI, 296 Seiten. 1973. DM 27,-

Vol. 334: F. Schweiger, The Metrical Theory of Jacobi-Perron Algorithm. V, 111 pages. 1973. DM 16,-

Vol. 335: H. Huck, R. Roitzsch, U. Simon, W. Vortisch, R. Walden, B. Wegner und W. Wendland, Beweismethoden der Differentialgeometrie im Großen. IX, 159 Seiten. 1973. DM 20,-

Vol. 336: L'Analyse Harmonique dans le Domaine Complexe. Edité par E. J. Akutowicz. VIII, 169 pages. 1973. DM 20,-

Vol. 337: Cambridge Summer School in Mathematical Logic. Edited by A. R. D. Mathias and H. Rogers. IX, 660 pages. 1973. DM 46,-

Vol: 338: J. Lindenstrauss and L. Tzafriri, Classical Banach Spaces. IX, 243 pages. 1973. DM 24,-

Vol. 339: G. Kempf, F. Knudsen, D. Mumford and B. Saint-Donat, Toroidal Embeddings I. VIII, 209 pages. 1973. DM 22,-

Vol. 340: Groupes de Monodromie en Géométrie Algébrique. (SGA 7 II). Par P. Deligne et N. Katz. X, 438 pages. 1973. DM 44,-

Vol. 341: Algebraic K-Theory I, Higher K-Theories. Edited by H. Bass. XV, 335 pages. 1973. DM 29,-

Vol. 342: Algebraic K-Theory II, "Classical" Algebraic K-Theory, and Connections with Arithmetic. Edited by H. Bass. XV, 527 pages. 1973. DM 40,-

Vol. 343: Algebraic K-Theory III, Hermitian K-Theory and Geometric Applications. Edited by H. Bass. XV, 572 pages. 1973. DM 40,-

Vol. 344: A. S. Troelstra (Editor), Metamathematical Investigation of Intuitionistic Arithmetic and Analysis. XVII, 485 pages. 1973. DM 38,-

Vol. 345: Proceedings of a Conference on Operator Theory. Edited by P. A. Fillmore. VI, 228 pages. 1973. DM 22,-

Vol. 346: Fučík et al., Spectral Analysis of Nonlinear Operators. II, 287 pages. 1973. DM 26,-

Vol. 347: J. M. Boardman and R. M. Vogt, Homotopy Invariant Algebraic Structures on Topological Spaces. X, 257 pages. 1973. DM 24,-

Vol. 348: A. M. Mathai and R. K. Saxena, Generalized Hypergeometric Functions with Applications in Statistics and Physical Sciences. VII, 314 pages. 1973. DM 26,-

Vol. 349: Modular Functions of One Variable II. Edited by W. Kuyk and P. Deligne. V, 598 pages. 1973. DM 38,-

Vol. 350: Modular Functions of One Variable III. Edited by W. Kuyk and J.-P. Serre. V, 350 pages. 1973. DM 26,-

Vol. 351: H. Tachikawa, Quasi-Frobenius Rings and Generalizations. XI, 172 pages. 1973. DM 20,-

Vol. 352: J. D. Fay, Theta Functions on Riemann Surfaces. V, 137 pages. 1973. DM 18,-

Vol. 353: Proceedings of the Conference on Orders, Group Rings and Related Topics. Organized by J. S. Hsia, M. L. Madan and T. G. Ralley. X, 224 pages. 1973. DM 22,-

Vol. 354: K. J. Devlin, Aspects of Constructibility. XII, 240 pages. 1973. DM 24,-

Vol. 355: M. Sion, A Theory of Semigroup Valued Measures. V, 140 pages. 1973. DM 18,-

Vol. 356: W. L. J. van der Kallen, Infinitesimally Central-Extensions of Chevalley Groups. VII, 147 pages. 1973. DM 18,-

Vol. 357: W. Borho, P. Gabriel und R. Rentschler, Primideale in Einhüllenden auflösbarer Lie-Algebren. V, 182 Seiten. 1973. DM 20,-

Vol. 358: F. L. Williams, Tensor Products of Principal Series Representations. VI, 132 pages. 1973. DM 18,-

Vol. 359: U. Stammbach, Homology in Group Theory. VIII, 183 pages. 1973. DM 20,-

Vol. 360: W. J. Padgett and R. L. Taylor, Laws of Large Numbers for Normed Linear Spaces and Certain Fréchet Spaces. VI, 111 pages. 1973. DM 18,-

Vol. 361: J. W. Schutz, Foundations of Special Relativity: Kinematic Axioms for Minkowski Space-Time. XX, 314 pages. 1973. DM 26,-

Vol. 362: Proceedings of the Conference on Numerical Solution of Ordinary Differential Equations. Edited by D. Bettis. VIII, 490 pages. 1974. DM 34,-

Vol. 363: Conference on the Numerical Solution of Differential Equations. Edited by G. A. Watson. IX, 221 pages. 1974. DM 20,-

Vol. 364: Proceedings on Infinite Dimensional Holomorphy. Edited by T. L. Hayden and T. J. Suffridge. VII, 212 pages. 1974. DM 20,-

Vol. 365: R. P. Gilbert, Constructive Methods for Elliptic Equations. VII, 397 pages. 1974. DM 26,-

Vol. 366: R. Steinberg, Conjugacy Classes in Algebraic Groups (Notes by V. V. Deodhar). VI, 159 pages. 1974. DM 18,-

Vol. 367: K. Langmann und W. Lütkebohmert, Cousinverteilungen und Fortsetzungssätze. VI, 151 Seiten. 1974. DM 16,-

Vol. 368: R. J. Milgram, Unstable Homotopy from the Stable Point of View. V, 109 pages. 1974. DM 16,-

Vol. 369: Victoria Symposium on Nonstandard Analysis. Edited by A. Hurd and P. Loeb. XVIII, 339 pages. 1974. DM 26,-